Lambert M. Surhone, Mariam T. Tennoe,
Susan F. Henssonow (Ed.)

Sony Ericsson Xperia X10 Mini Pro

Lambert M. Surhone, Mariam T. Tennoe,
Susan F. Henssonow (Ed.)

Sony Ericsson Xperia X10 Mini Pro

Smartphone, Xperia, Android (Operating System)

Betascript Publishing

Imprint

All parts of this book are extracted from Wikipedia, the free encyclopedia (www.wikipedia.org).

You can get detailed informations about the authors of this collection of articles at the end of this book. The editors (Ed.) of this book are no authors. They have not modified or extended the original texts.

Pictures published in this book can be under different licences than the GNU Free Documentation License. You can get detailed informations about the authors and licences of pictures at the end of this book.

The content of this book was generated collaboratively by volunteers. Please be advised that nothing found here has necessarily been reviewed by people with the expertise required to provide you with complete, accurate or reliable information. Some information in this book maybe misleading or wrong. The Publisher does not guarantee the validity of the information found here. If you need specific advice (f.e. in fields of medical, legal, financial, or risk management questions) please contact a professional who is licensed or knowledgeable in that area.

Cover image: www.ingimage.com
Concerning the licence of the cover image please contact ingimage.

Contact:
VDM Publishing House Ltd.,17 Rue Meldrum, Beau Bassin,1713-01 Mauritius
Email: info@vdm-publishing-house.com
Website: www.vdm-publishing-house.com

Published in 2010
Printed in: U.S.A., U.K., Germany. This book was not produced in Mauritius.

ISBN: 978-613-3-69286-2

Contents

Articles

Sony Ericsson Xperia X10 Mini Pro 1
Sony Ericsson Xperia X10 Mini 2
Smartphone 4
Xperia 8
Android (operating system) 11
GPRS 31
UMTS 37
HSDPA 47
Sony Ericsson XPERIA X10 Mini Pro 51
Bluetooth 52
Sony Ericsson 71

References

Article Sources and Contributors 80
Image Sources, Licenses and Contributors 83

Article Licenses

License 84

Sony Ericsson Xperia X10 Mini Pro

The **Sony Ericsson Xperia X10 Mini Pro** is a mobile telephone released by Sony Ericsson on the 24 May 2010.

The device is an upgrade of the similar X10 Mini with many of the internal specifications being identical. The major differences between the X10 Mini and X10 Mini Pro are a slide-out full QWERTY Keyboard, and the Pro having slightly larger dimensions (3.5 × 2.0 × 0.7 inches opposed to 3.3 × 2.0 × 0.6 inches).

The X10 Mini and X10 Mini Pro are designed to look similar and share functionality with the larger Xperia X10, but are internally very different devices. The X10 Mini and X10 Mini Pro lack Sony Ericsson's "Mediascape" media-management software, but include "Timescape" as well as the proprietary "Rachael" UI.

The X10 Mini Pro (as well as the X10 and X10 Mini) runs on Android 1.6, with an update to 2.1 being rolled out from Sunday 31st October 2010, to Tuesday 30th November 2010.

References

- Sony Ericsson X10 Mini Official Page [1]
- Sony Ericsson X10 Mini Pro Official Page [2]

References

[1] http://www.sonyericsson.com/cws/products/mobilephones/overview/xperiax10mini?cc=gb&lc=en#view=overview
[2] http://www.sonyericsson.com/cws/products/mobilephones/overview/xperiax10minipro?cc=gb&lc=en

Sony Ericsson Xperia X10 Mini

Manufacturer	Sony Ericsson
Available	Q2 2010[1]
Screen	240 x 320 pixels (QVGA) 16M color TFT
Camera	5 MP with Auto focus, Geo tagging,and Touch focus
Operating system	Android 2.1 Android 1.6 (Shipped)[2]
Input	Touchscreen
CPU	600 MHz Qualcomm MSM7227
Memory	256 MB RAM
Networks	2G Quad-band GSM/GPRS Class 10 (4+1/3+2 slots), 32 - 48 kbit/s/EDGE: 850/900/1800/1900 MHz 3G Tri-band UMTS/HSDPA/HSUPA: 850/1900/2100/900 MHz
Connectivity	Bluetooth 2.0 with A2DP microUSB 2.0 3.5mm audio jack aGPS Wi-Fi 802.11 b/g no IR
Battery	Standard battery: Li-Po 1500 mAh (BST-41). Extended batteries available [3]
Physical size	83.0 x 50.0 x 16.0 mm
Weight	88 g with battery
Series	Sony Ericsson Xperia series

The **Sony Ericsson XPERIA X10 Mini (E10i)** is a smartphone by Sony Ericsson in the Xperia series. It is the second Sony Ericsson smartphone to run the Android operating system[2] and also the smallest Android handset to date. It origenly started with the Android 1.6 operating system, but has recently changed to Android OS 2.1.

The X10 Mini was first revealed on 14 February 2010.[1] Apart from the physical dimensions, the phone also differs feature-wise as compared to the X10. Text input is done via a virtual keypad and not a Qwerty keyboard like on the Xperia X10. It comes with a 5MP camera, differing to the X10's 8MP. It also comes with a T9 dictionary which makes it easy to use the keypad. Also, there are tabs for easily accessing symbols and numbers.[4]

Awards

"2010 red dot product design award" [5]

"European Mobile Phone 2010-2011" [6]

See also

- Sony Ericsson XPERIA X10 Mini Pro
- List of Android devices

References

[1] "Get Compact and Clever with Sony Ericsson Xperia™ X10 mini and Xperia™ X10mini pro" (http://www.sonyericsson.com/cws/corporate/press/pressreleases/pressreleasedetails/sonyericssonx10miniandminifinal-20100214). *Sony Ericsson*. 14 February 2010. .

[2] "Sony Ericsson Xperia X10 Mini / Mini Pro review" (http://www.engadget.com/2010/07/08/sony-ericsson-xperia-x10-mini-mini-pro-review/). *Engadget*. 8 July 2010. . Retrieved November 20, 2009.

[3] "Mugen Power Batteries Releases 3600mAh Extended Battery for Sony-Ericsson Xperia X10" (http://www.prweb.com/releases/2010/05/prweb3982044.htm). PRWeb.com. 12 May 2010. .

[4] Xperia X10 mini: Reviewed (http://techviewz.org/2010/09/xperia-x10-mini-reviewed.html)

[5] ""Compact and clever"" (http://www.sonyericsson.com/cws/companyandpress/aboutus/awards/award/x10miniaward?cc=gb&lc=en). *Sony Ericsson*. .

[6] ""Xperia™ X10 mini - European Mobile Phone 2010-2011"" (http://www.sonyericsson.com/cws/companyandpress/aboutus/awards/award/x10minieisaaward?cc=gb&lc=en). *Sony Ericsson*. .

External links

- Xperia X10 Mini website (http://www.sonyericsson.com/cws/products/mobilephones/overview/xperiax10mini)
- XDA Developers X10 Mini Wiki (http://forum.xda-developers.com/wiki/index.php?title=Xperia_X10_Mini)

Smartphone

A **smartphone** is a mobile phone that offers more advanced computing ability and connectivity than a contemporary basic feature phone.[2] Smartphones and feature phones may be thought of as handheld computers integrated within a mobile telephone, but while most feature phones are able to run applications based on platforms such as Java ME,[3] a smartphone allows the user to install and run more advanced applications based on a specific platform. Smartphones run complete operating system software providing a platform for application developers.[4] A smartphone can be considered as a Personal Pocket Computer (PPC) with mobile phone functions, because these devices are mainly computers, although quite smaller than a desktop computer (DC). Additionally a PPC (Personal Pocket Computer) is more personal than a DC (desktop computer).

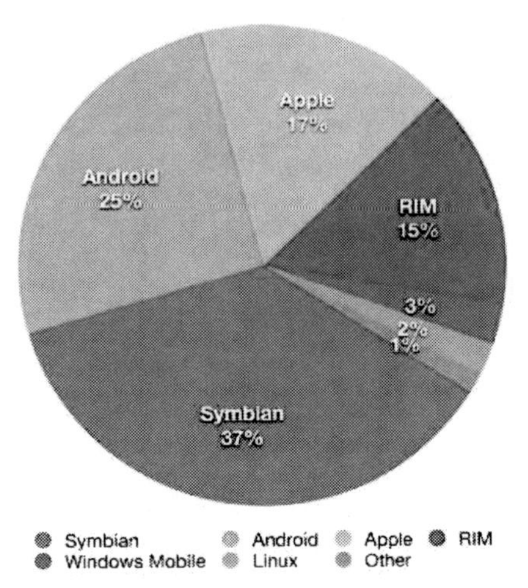

Share of 2010 Q3 smartphone sales to end users by operating system, according to Gartner.[1]

Growth in demand for advanced mobile devices boasting powerful processors, abundant memory, larger screens, and open operating systems has outpaced the rest of the mobile phone market for several years.[5] According to a study by ComScore, over 45.5 million people in the United States owned smartphones in 2010 and it is the fastest growing segment of the mobile phone market, which comprised 234 million subscribers in the United States.[6] Despite the large increase in smartphone sales in the last few years, smartphone shipments only make up 20% of total handset shipments, as of the first half of 2010.[7]

History

Early years

The first smartphone was called Simon; it was designed by IBM in 1992 and shown as a concept product[8] that year at COMDEX, the computer industry trade show held in Las Vegas, Nevada. It was released to the public in 1993 and sold by BellSouth. Besides being a mobile phone, it also contained a calendar, address book, world clock, calculator, note pad, e-mail, send and receive fax, and games. It had no physical buttons to dial with. Instead customers used a touchscreen to select telephone numbers with a finger or create facsimiles and memos with an optional stylus. Text was entered with a unique on-screen "predictive" keyboard. By today's standards, the Simon would be a fairly low-end product; however, its feature set at the time was highly advanced.

The Nokia Communicator line was the first of Nokia's smartphones starting with the Nokia 9000, released in 1996. This distinctive palmtop computer style smartphone was the result of a collaborative effort of an early successful and costly personal digital assistant (PDA) by Hewlett Packard combined with Nokia's bestselling phone around that time, and early prototype models had the two devices fixed via a hinge. The Nokia 9210 was the first color screen Communicator model which was the first true smartphone with an open operating system; the 9500 Communicator was also Nokia's first cameraphone Communicator and Nokia's first WiFi phone. The 9300 Communicator was the third dimensional shift into a smaller form factor, and the latest E90 Communicator includes GPS. The Nokia

Communicator model is remarkable for also having been the most costly phone model sold by a major brand for almost the full life of the model series, costing easily 20% and sometimes 40% more than the next most expensive smartphone by any major producer.

In 1997 Ericsson released the concept phone GS88,[9] [10] the first device labelled as 'smartphone'.[11]

Rise of Symbian, Windows Mobile, and BlackBerry

In 2000 Ericsson released the touchscreen smartphone R380, the first device to use the new Symbian OS.[12] It was followed up by P800 in 2002, the first camera smartphone.[13]

In 2001 Microsoft announced its Windows CE Pocket PC OS would be offered as "Microsoft Windows Powered Smartphone 2002."[14] Microsoft originally defined its Windows Smartphone products as lacking a touchscreen and offering a lower screen resolution compared to its sibling Pocket PC devices.

In early 2002 Handspring released the Palm OS Treo smartphone, utilizing a full keyboard that combined wireless web browsing, email, calendar, and contact organizer with mobile third-party applications that could be downloaded or synced with a computer.[15]

In 2002 RIM released the first BlackBerry which was the first smartphone optimized for wireless email use and had achieved a total customer base of 32 million subscribers by December 2009.[16]

In 2007 Nokia launched the Nokia N95 which integrated a wide range of features into a consumer-oriented smartphone: GPS, a 5 megapixel camera with autofocus and LED flash, 3G and wi-fi connectivity and TV-out. In the next few years these features would become standard on high-end smartphones.

Rise of the iPhone and Android

Later in 2007, Apple Inc. introduced its first iPhone. It was initially costly, priced at $500 for the cheaper of two models on top of a two year contract. It was one of the first smartphones to be mainly controlled through its touchscreen, the others being the LG Prada and the HTC Touch (also released in 2007). It was the first mobile phone to use a multi-touch interface. It featured a web browser that was much better than its competitors - *Ars Technica* described it as "far superior to anything that we had ever used prior."[17] At the time of the launch of the iPhone it was arguable whether it was actually a smartphone as the first generation lacked the ability to officially use third-party applications.[18] A process called jailbreaking emerged quickly to provide unofficial third-party applications. Steve Jobs publicly stated that the iPhone lacked 3G support due to the immaturity, power use, and physical size requirements of 3G chipsets at the time.[19] However, it has been rumored that the CDMA2000 Network Providers (Verizon, Sprint) refused to allow the iPhone on their network because Jobs wanted total control of the application store associated with the iPhone.

Android, a cross platform OS for smartphones was released in 2008. Android is an open source platform backed by Google, along with major hardware and software developers (such as Intel, HTC, ARM, Motorola and Samsung, to name a few), that form the Open Handset Alliance.[20] The first phone to use the Android OS was the HTC Dream, branded for distribution by T-Mobile as the G1.[21] The software suite included on the phone consists of integration with Google's proprietary applications, such as Maps, Calendar, and Gmail, and a full HTML web browser. Third-party apps are available via the Android Market, including both free and paid apps.[22]

In July 2008, Apple introduced its second generation iPhone which had a lower list price and 3G support. It also created the App Store with both free and paid applications. The App Store can deliver smartphone applications developed by third parties directly to the iPhone or iPod Touch over wifi or cellular network without using a PC to download. The App Store has been a huge success for Apple and by April 2010 hosted more than 185,000 applications.[23] The App Store hit three billion application downloads in early January 2010.[24]

Resurgence of Nokia

In 2010 Nokia have been planning a fightback in the smartphone market with the Nokia N8 smartphone, the first device to use the new Symbian^3 OS.[25] The Nokia N8 smartphone will shortly be followed by the Nokia C7 smartphone and the Nokia E7 smartphone.[26] [27]

Other platforms are able to download apps from any website, rather than only from a single app store; however, other companies have more recently launched their own app stores. RIM launched its app store, BlackBerry App World, in April 2009. Nokia launched its Ovi Store in May 2009. Palm launched its Palm App Catalog in June 2009. Microsoft launched its Windows Marketplace for Mobile in October 2009.

In January 2010, Google launched Nexus One using its Android OS. Although Android OS has multi-touch capabilities, Google initially removed that feature from Nexus One,[28] but it was added through a firmware update on February 2, 2010.[29]

Operating systems

Many mobile operating systems exist and are in use. As of 2010, the biggest selling smartphone operating system is Symbian OS,[30] Symbian's smaller rivals include Android, Blackberry OS, iOS and the Windows Phone OS. Several mobile operating systems including Android and iOS are based on Linux and Unix.

Smartbook

A smartbook is a concept of a mobile device that falls between smartphones and netbooks, delivering features typically found in smartphones (always on, all-day battery life, 3G connectivity, GPS)[31] in a slightly larger device with a full keyboard. Smartbooks will tend to be designed to work with online applications.[32]

Smartbooks use the ARM processor, which gives them much greater battery life than a netbook which uses a traditional Intel x86 processor.[33] They are likely to be sold initially through mobile network operators, like mobile phones are today, along with a wireless data plan.[34]

Open source development

The open source culture has penetrated the smartphone market in several ways. There have been attempts to open source both hardware and software of smartphones. The most notable project from open hardware development is most likely the Neo FreeRunner smartphone developed by Openmoko. Lately, the Google Android OS is a popular open source mobile operating system. Nokia has an initiative around Symbian too, which open-sourced all Symbian smartphone code in February 2010.[35] Nokia has developed a GNU/Linux-based open source system Maemo. Later, Maemo was merged with Intel's project Moblin to form MeeGo operating system.

See also

- Comparison of smartphones
- Blackberry thumb
- Camera phone
- e-book reader
- Energy harvesting
- Flexible keyboard
- HDMI
- Information appliance
- Tablet PC
- mAh, to measure battery capacity
- Microbrowser
- Memory card
- Text-to-speech
- Videophones
- PEC-Phone (Pocket Entertainment Computer-Phone)
- List of digital distribution platforms for mobile devices (app stores)
- Personal Handy-phone System
- Pay As You Go (phone)
- SIM card
- Mobile broadband
- Tethering
- Mobile internet device (MID)
- Smartbook
- Netbook
- Laptop
- iPhone

References

[1] Gartner (10 November 2010). "Gartner Says Worldwide Mobile Phone Sales Grew 35 Percent in Third Quarter 2010; Smartphone Sales Increased 96 Percent" (http://www.gartner.com/it/page.jsp?id=1466313). Press release. . Retrieved 18 November 2010.

[2] Andrew Nusca (20 August 2009). "Smartphone vs. feature phone arms race heats up; which did you buy?" (http://www.zdnet.com/blog/gadgetreviews/smartphone-vs-feature-phone-arms-race-heats-up-which-did-you-buy/6836). ZDNet. .

[3] "Feature Phone" (http://www.phonescoop.com/glossary/term.php?gid=310). Phone Scoop. . Retrieved 9 May 2010.

[4] "Smartphone definition from PC Magazine Encyclopedia" (http://www.pcmag.com/encyclopedia_term/0,2542,t=Smartphone&i=51537,00.asp). *PC Magazine*. . Retrieved 13 May 2010.

[5] "Smart phones: how to stay clever in downturn" (http://www.deloitte.co.uk/TMTPredictions/telecommunications/Smartphones-clever-in-downturn.cfm). *Deloitte Telecommunications Predictions*. .

[6] "Android Phones Steal Market Share" (http://bmighty.informationweek.com/mobile/showArticle.jhtml?articleID=224201881). .

[7] "100 Million Club - H1 2010" (http://www.visionmobile.com/blog/2010/10/smart-feature-phones-the-unbalanced-equation-100-million-club-series/). .

[8] Schneidawind, J: "Big Blue unveiling", *USA Today*, November 23, 1992, page 2B

[9] "Ericsson GS88 Preview" (http://pws.prserv.net/Eri_no_moto/GS88_Preview.htm). *Eri-no-moto*. . Retrieved 19 September 2010.

[10] "History" (http://www.stockholmsmartphone.org/history/). Stockholm Smartphone. . Retrieved 19 September 2010.

[11] "Ericsson GS88 box" (http://www.stockholmsmartphone.org/wp-content/uploads/penelope-box.jpg). . Retrieved 19 September 2010.

[12] "Symbian Device - The OS Evolution" (http://www.i-symbian.com/wp-content/uploads/2009/11/Symbian_Evolution.pdf). Independent Symbian Blog. . Retrieved 12 August 2010.

[13] "P800" (http://www.allaboutsymbian.com/features/item/Sony_Ericsson_P800.php). All About Symbian. . Retrieved 12 August 2010.

[14] Windows Powered Smartphone (http://www.microsoft.com/presspass/features/2002/Jan02/01-08msces.mspx)

[15] Handspring's Breakthrough Hybrid (http://www.businessweek.com/bwdaily/dnflash/nov2001/nf20011129_0157.htm)

[16] BlackBerry Users Call For RIM To Rethink Service (http://www.crn.com/mobile/222002587;jsessionid=DTFGA1NIW4DPVQE1GHRSKH4ATMY32JVN)

[17] "iPhone in depth: the Ars review" (http://arstechnica.com/apple/reviews/2007/07/iphone-review.ars/6). *ArsTechnica*. Condé Nast. 9 July 2007. p. 6. . Retrieved 3 August 2010.

[18] "The iPhone is not a smartphone" (http://www.engadget.com/2007/01/09/the-iphone-is-not-a-smartphone/). Engadget. 9 January 2007. . Retrieved 11 July 2010.

[19] "No 3G on the iPhone, but why? A battery life analysis" (http://www.anandtech.com/show/2274). *AnandTech*. 13 July 2007. p. 1. .

[20] http://www.openhandsetalliance.com/oha_members.html

[21] http://www.t-mobileg1.com/?WT.srch=1&WT.mc_id=273m1&WT.z=p137999826

[22] Paid apps still coming to Android Market in Q1 '09, US and UK rollout first (http://www.engadget.com/2008/12/31/paid-apps-still-coming-to-android-market-in-q1-09-us-and-uk-ro)

[23] "iPhone App Reviews by Experts at PCWorld - PCWorld" (http://www.pcworld.com/appguide/index.html). PCWorld. . Retrieved 25 February 2010.

[24] "Apple's App Store Downloads Top Three Billion" (http://www.apple.com/pr/library/2010/01/05appstore.html). Apple. 5 January 2010. . Retrieved 19 January 2010.

[25] "Nokia N8 smartphone official site" (http://www.nokia.co.uk/n8). .

[26] "Nokia C7 smartphone official site" (http://www.nokia.co.uk/C7). .

[27] "Nokia E7 smartphone official site" (http://www.nokia.co.uk/E7). .

[28] Weighing Nexus over iPhone - a practical review for the everyday user! (http://techietrick.blogspot.com/2010/01/weighing-nexus-over-iphone-practical.html)
[29] Nexus One gets a software update, enables multitouch (updated with video!) (http://www.engadget.com/2010/02/02/nexus-one-gets-a-software-update-enables-multitouch)
[30] "Symbian official site" (http://www.symbian.org/news-and-media/2010/09/14/symbian-welcomes-new-nokia-smartphones-symbian3-family). .
[31] http://www.eetimes.eu/design/217700855
[32] Schofield, Jack (29 July 2009). "The smartbook has been waiting 28 years to be the next best thing" (http://www.guardian.co.uk/technology/2009/jul/29/smartbooks-netbooks). *The Guardian* (London). . Retrieved 23 May 2010.
[33] Scott Stein (10 January 2010). "CES: What, exactly, is a smartbook? Highlights from the show floor" (http://ces.cnet.com/8301-31045_1-10431884-269.html). cnet. . Retrieved 2010-01-11.
[34] Ganapat, Priya (2008-12-15). " The Next Netbook Trend: Cellphone-Like Contract Deals (http://blog.wired.com/gadgets/2008/12/next-time-you-s.html)" – Wired News.
[35] Symbian OS, Now Fully Open Source (http://www.watblog.com/2010/02/06/symbian-os-now-fully-open-source)

External links

- freesmartphone.org (http://www.freesmartphone.org) - a collaboration platform for open source smartphones
- Smart mobile phone emulator (http://www.gameldigital.com/net/iphone/iphone-en-wikipedia.php)
- The Smartphone Genome Project (http://www.wikismartphones.org/index.php/Main_Page)
- Defining the Smartphone, part 1 (http://www.allaboutsymbian.com/features/item/Defining_the_Smartphone.php), part 2 (http://www.allaboutsymbian.com/features/item/Spy_versus_Spy_No_Smartphone_versus_Smartphone.php) by Steve Litchfield on July 16th 2010, various definitions are treated

pfl:Smartphone

Xperia

Xperia is the name of Sony Ericsson's range of high-end smartphones. Five members of the product family have been released:

- Xperia X1 - released in October 2008, runs on Windows Mobile 6.1
- Xperia X2 - released in September 2009, runs on Windows Mobile 6.5[1]
- Xperia X10 - released in March 2010, runs on Android 1.6[2] / 2.1[3]
- Xperia X10 Mini - released in May 2010, runs on Android 1.6[4] / 2.1[3]
- Xperia X10 Mini Pro - released in May 2010, runs Android 1.6[5] / 2.1[3]
- Xperia X8 - released in Q4 2010, running Android 1.6[6] / 2.1[3]

Initial Software Issues

The initial 3UK (and possibly other Mobile Operators) release of firmware contains several bugs - Some users report WiFi reception is very poor[7] , and also if the user has a PIN code active on their USIM, the phone can PUK-lock the user's USIM without warning. Once the USIM is PUK-locked, there is no way to unlock the USIM in the Xperia handset - the USIM must be placed in another mobile phone to apply the PUK unlock code.

Constant Delays in Updates

Promised update to Android 2.1 and much needed patches for critical software bugs has its release date pushed back time and time again.[8]

Build Quality Issues

X1 and X2 are subject to cracks that appear without reason. Warranty pertaining to the issue is honoured in selected countries but simply denied and ignored by customer service in the others.[9] [10]

Upcoming models

- Xperia Pureness, a translucent phone without camera that will be sold by selected retailers in selected cities.[11]

X Series: Xperia phones

Phone model	Screen type	Screen resolution	Released	Technology/Frequency	Form factor	Camera	Carrier(s)	OS
Sony Ericsson Xperia X1	Color	WVGA	2008	GSM, UMTS	Slider	3.2 Megapixel	Rogers Wireless in Canada but no carriers in US	Windows Mobile 6.1
Sony Ericsson Xperia X2	Color	WVGA	2009	GSM, UMTS	Slider	8.1 Megapixel	Unknown	Windows Mobile 6.5
Sony Ericsson Xperia X8	Color	HVGA	Q4 2010	GSM, UMTS	Touchscreen	3 Megapixel	Unknown	Android 1.6 / 2.1
Sony Ericsson Xperia X10	Color	FWVGA	2010[12] [13]	GSM, UMTS	Touchscreen	8.1 Megapixel	Rogers Wireless in Canada and AT&T in the US	Android 1.6 / 2.1
Sony Ericsson Xperia Pureness	Monochrome	Unknown	2009[14]	GSM, UMTS	Candybar	No	Unknown	A200 Platform
Sony Ericsson Xperia X10 Mini	Color	QVGA	2010[15]	GSM, UMTS	Touchscreen	5 Megapixel	Rogers Wireless in Canada	Android 1.6 / 2.1
Sony Ericsson Xperia X10 Mini Pro	Color	QVGA	2010[15]	GSM, UMTS	Slider	5 Megapixel	Unknown	Android 1.6 / 2.1

See also

- Technological convergence

References

[1] Sony Ericsson Unveils XPERIA X2 Smartphone (http://www.cbronline.com/news/sony_ericsson_unveils_xperia_x2_smartphone_090903)

[2] "Sony Ericsson XPERIA X10 - Full phone specifications" (http://www.gsmarena.com/sony_ericsson_xperia_x10-2964.php). GSMArena.com. . Retrieved 27 June 2010.

[3] "X8 Announcement and XPERIA Updates" (http://blogs.sonyericsson.com/products/2010/06/16/x8-announcement-and-xperia-updates/). sonyericsson.com. . Retrieved 21 July 2010.

[4] "Sony Ericsson XPERIA X10 mini - Full phone specifications" (http://www.gsmarena.com/sony_ericsson_xperia_x10_mini-3125.php). GSMArena.com. . Retrieved 27 June 2010.

[5] "Sony Ericsson XPERIA X10 mini pro - Full phone specifications" (http://www.gsmarena.com/sony_ericsson_xperia_x10_mini_pro-3147.php). GSMArena.com. . Retrieved 27 June 2010.

[6] "Sony Ericsson XPERIA X8 - Full phone specifications" (http://www.gsmarena.com/sony_ericsson_xperia_x8-3403.php). GSMArena.com. . Retrieved 27 June 2010.

[7] Xperia X10 WiFi bug found (http://blog.se-nse.net/xperia-x10-wifi-bug-cause-found/)

[8] [http://blogs.sonyericsson.com/products/2010/09/23/update-on-updates/

[9] http://forum.xda-developers.com/showthread.php?t=447211
[10] http://forum.xda-developers.com/showthread.php?t=691611
[11] Xperia Pureness - genomskinlig Sony Ericsson (http://www.mobil.se/ArticlePages/200909/04/20090904095516_MOB_Administratorer004/20090904095516_MOB_Administratorer004.dbp.asp)
[12] Sony Ericsson XPERIA™ X10 introduces an open and integrated world of social media, communication and entertainment (http://www.sonyericsson.com/cws/corporate/press/pressreleases/pressreleasedetails/xperiax10pressreleasefinal-20091103)
[13] Sony Ericsson unveils Xperia X10 Andriod handset (http://www.cbronline.com/news/sony_ericsson_unveils_xperia_x10_andriod_handset_091103)
[14] Sony Ericsson shows new Xperia Pureness at London brand event (http://www.sonyericsson.com/cws/corporate/press/pressreleases/pressreleasedetails/websitecopy-20090903)
[15] http://www.sonyericsson.com/cws/corporate/press/pressreleases/pressreleasedetails/sonyericssonx10miniandminifinal-20100214

Android (operating system)

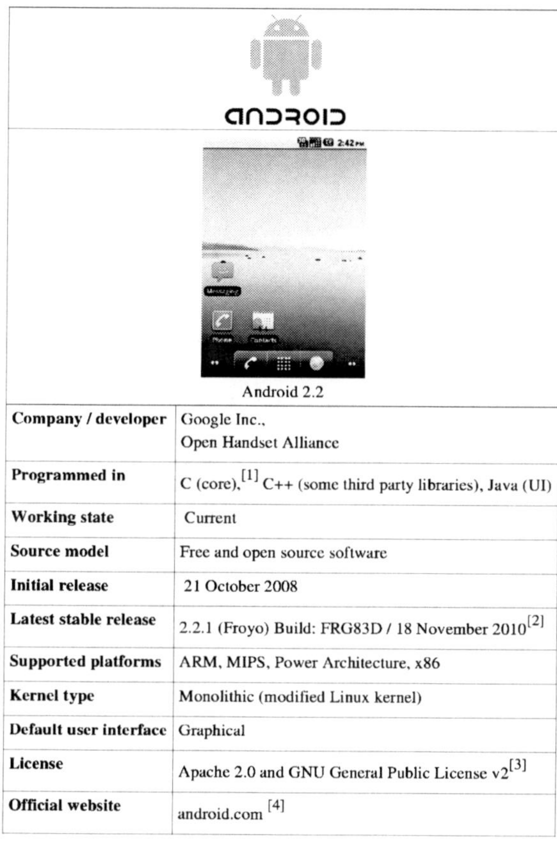

Android 2.2

Company / developer	Google Inc., Open Handset Alliance
Programmed in	C (core),[1] C++ (some third party libraries), Java (UI)
Working state	Current
Source model	Free and open source software
Initial release	21 October 2008
Latest stable release	2.2.1 (Froyo) Build: FRG83D / 18 November 2010[2]
Supported platforms	ARM, MIPS, Power Architecture, x86
Kernel type	Monolithic (modified Linux kernel)
Default user interface	Graphical
License	Apache 2.0 and GNU General Public License v2[3]
Official website	android.com [4]

Android is a mobile operating system initially developed by Android Inc., a firm purchased by Google in 2005.[5] Android is based upon a modified version of the Linux kernel. Google and other members of the Open Handset Alliance collaborated to develop and release Android to the world.[6] [7] The Android Open Source Project (AOSP) is tasked with the maintenance and further development of Android.[8] Unit sales for Android OS smartphones ranked first among all smartphone OS handsets sold in the U.S. in the second and third quarters of 2010,[9] [10] [11] with a third quarter market share of 43.6%.[12]

Android has a large community of developers writing application programs *("apps")* that extend the functionality of the devices. There are currently over 100,000 apps available for Android.[13] [14] Android Market is the online app store run by Google, though apps can be downloaded from third party sites (except on AT&T, which disallows this). Developers write in the Java language, controlling the device via Google-developed Java libraries.[15]

The unveiling of the Android distribution on 5 November 2007 was announced with the founding of the Open Handset Alliance, a consortium of 78 hardware, software, and telecom companies devoted to advancing open standards for mobile devices.[16] [17] Google released most of the Android code under the Apache License, a free software and open source license.[18]

The Android operating system software stack consists of Java applications running on a Java based object oriented application framework on top of Java core libraries running on a Dalvik virtual machine featuring JIT compilation. Libraries written in C include the surface manager, OpenCore[19] media framework, SQLite relational database management system, OpenGL ES 2.0 3D graphics API, WebKit layout engine, SGL graphics engine, SSL, and Bionic libc. The Android operating system consists of 12 million lines of code including 3 million lines of XML, 2.8 million lines of C, 2.1 million lines of Java, and 1.75 million lines of C++.[20]

History

Acquisition by Google

In July 2005, Google acquired Android, Inc., a small startup company based in Palo Alto, California, USA.[21] Android's co-founders who went to work at Google included Andy Rubin (co-founder of Danger),[22] Rich Miner (co-founder of Wildfire Communications, Inc.),[23] Nick Sears (once VP at T-Mobile),[24] and Chris White (headed design and interface development at WebTV).[25] At the time, little was known about the functions of Android, Inc. other than that they made software for mobile phones.[21] This began rumors that Google was planning to enter the mobile phone market.

At Google, the team led by Rubin developed a mobile device platform powered by the Linux kernel which they marketed to handset makers and carriers on the premise of providing a flexible, upgradable system. It was reported that Google had already lined up a series of hardware component and software partners and signaled to carriers that it was open to various degrees of cooperation on their part.[26] [27] [28] More speculation that Google would be entering the mobile-phone market came in December 2006.[29] Reports from the BBC and *The Wall Street Journal* noted that Google wanted its search and applications on mobile phones and it was working hard to deliver that. Print and online media outlets soon reported rumors that Google was developing a Google-branded handset.[30] More speculation followed reporting that as Google was defining technical specifications, it was showing prototypes to cell phone manufacturers and network operators.

In September 2007, *InformationWeek* covered an Evalueserve study reporting that Google had filed several patent applications in the area of mobile telephony.[31] [32]

Open Handset Alliance

"Today's announcement is more ambitious than any single 'Google Phone' that the press has been speculating about over the past few weeks. Our vision is that the powerful platform we're unveiling will power thousands of different phone models."

Eric Schmidt, *Google Chairman/CEO*[6]

On the 5th of November 2007, the Open Handset Alliance, a consortium of several companies which include Texas Instruments, Broadcom Corporation, Google, HTC, Intel, LG, Marvell Technology Group, Motorola, Nvidia, Qualcomm, Samsung Electronics, Sprint Nextel and T-Mobile was unveiled with the goal to develop open standards for mobile devices.[6] Along with the formation of the Open Handset Alliance, the OHA also unveiled their first product, Android, a mobile device platform built on the Linux kernel version 2.6.[6]

On 9 December 2008, it was announced that 14 new members would be joining the Android Project, including PacketVideo, ARM Holdings, Atheros Communications, Asustek Computer Inc, Garmin Ltd, Softbank, Sony Ericsson, Toshiba Corp, and Vodafone Group Plc.[33] [34]

Licensing

With the exception of brief update periods, Android has been available under a free software / open source license since 21 October 2008. Google published the entire source code (including network and telephony stacks)[35] under an Apache License.[36]

The Apache License allows vendors to add proprietary extensions without submitting them back to the open source community.

Update history

Android has seen a number of updates since its original release. These updates to the base operating system typically fix bugs and add new features. Generally each update to the Android operating system is developed under a code name based on a dessert item.

<table>
<tr><td>1.1</td><td>Released 9 February 2009</td></tr>
<tr><td>1.5 (Cupcake)
Based on Linux Kernel 2.6.27</td><td>On 30 April 2009, the official 1.5 (Cupcake) update for Android was released.[37] [38] There were several new features and UI updates included in the 1.5 update:[39]
• Ability to record and watch videos through camcorder mode
• Uploading videos to YouTube and pictures to Picasa directly from the phone
• A new soft-keyboard with text-prediction
• Bluetooth A2DP and AVRCP support
• Ability to automatically connect to a Bluetooth headset within a certain distance
• New widgets and folders that can populate the Home screens
• Animated screen transitions</td></tr>
<tr><td>1.6 (Donut)
Based on Linux Kernel 2.6.29[40]</td><td>On 15 September 2009, the 1.6 (Donut) SDK was released.[41] [42] Included in the update were:[40]
• An improved Android Market experience
• An integrated camera, camcorder, and gallery interface
• Gallery now enables users to select multiple photos for deletion
• Updated Voice Search, with faster response and deeper integration with native applications, including the ability to dial contacts
• Updated search experience to allow searching bookmarks, history, contacts, and the web from the home screen
• Updated technology support for CDMA/EVDO, 802.1x, VPNs, and a text-to-speech engine
• Support for WVGA screen resolutions
• Speed improvements in searching and camera applications
• Gesture framework and GestureBuilder development tool
• Google free turn-by-turn navigation</td></tr>
<tr><td>2.0 / 2.1 (Eclair)
Based on Linux Kernel 2.6.29[43]</td><td>On 26 October 2009, the 2.0 (Eclair) SDK was released.[44] Among the changes were:[45]
• Optimized hardware speed
• Support for more screen sizes and resolutions
• Revamped UI
• New Browser UI and HTML5 support
• New contact lists
• Better contrast ratio for backgrounds
• Improved Google Maps 3.1.2
• Microsoft Exchange support
• Built in flash support for Camera
• Digital Zoom
• MotionEvent class enhanced to track multi-touch events[46]
• Improved virtual keyboard
• Bluetooth 2.1
• Live Wallpapers
The 2.0.1 SDK was released on 3 December 2009.[47]
The 2.1 SDK was released on 12 January 2010.[48]</td></tr>
</table>

2.2 (Froyo)[49] **Based on Linux Kernel 2.6.32**[50]	On 20 May 2010, the 2.2 (Froyo - Frozen Yogurt) SDK was released.[51] Changes included:[52] • General Android OS speed, memory, and performance optimizations[53] • Additional application speed improvements courtesy of JIT implementation[54] • Integration of Chrome's V8 JavaScript engine into the Browser application • Increased Microsoft Exchange support (security policies, auto-discovery, GAL look-up, calendar synchronization, remote wipe) • Improved application launcher with shortcuts to Phone and Browser applications • USB tethering and Wi-Fi hotspot functionality • Added an option to disable data access over mobile network • Updated Market application with batch and automatic update features[53] • Quick switching between multiple keyboard languages and their dictionaries • Voice dialing and contact sharing over Bluetooth • Support for numeric and alphanumeric passwords • Support for file upload fields in the Browser application[55] • Browser can now display animated GIFs (instead of just the first frame) • Support for installing applications to the expandable memory[56] • Adobe Flash 10.1 support[57]
2.3 (Gingerbread)[58] **Based on Linux Kernel 2.6.35.7**[50]	Confirmed new features of 2.3 (Gingerbread): • Support for WebM video playback[59] • Support for Near Field Communication [60] Unconfirmed new features: • Improved copy–paste functionalities[61] • Improved social networking features[62] • Android Market music store[63] • Media streaming from PC library[63] • Revamped UI[64] • Support for bigger screens with up to Wide XGA (1366×768) resolution[65] • New 3D Games support including new Marketplace area for gaming • Use of mksh for /system/bin/sh[66] • Support for video calls • Support for WebP image files • Support for Google TV
3.0 (Honeycomb)[67] [68]	Scheduled for early 2011 launch. Feature list started with features that won't make the cut-off for Gingerbread
? (Ice Cream)[69]	Supposed mid 2011 launch.[69]

Features

Current features and specifications:[70] [71] [72]

The Android Emulator default home screen (v1.5).

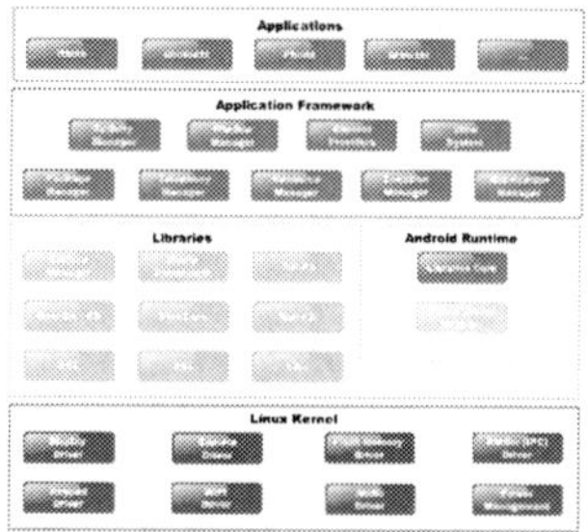

Architecture Diagram

Handset layouts	The platform is adaptable to larger, VGA, 2D graphics library, 3D graphics library based on OpenGL ES 2.0 specifications, and traditional smartphone layouts.
Storage	SQLite, a lightweight relational database, is used for data storage purposes
Connectivity	Android supports connectivity technologies including GSM/EDGE, IDEN, CDMA, EV-DO, UMTS, Bluetooth, Wi-Fi, LTE, and WiMAX.
Messaging	SMS and MMS are available forms of messaging, including threaded text messaging and now Android Cloud to Device Messaging Framework (C2DM) is also a part of Android Push Messaging service.
Web browser	The web browser available in Android is based on the open-source WebKit layout engine, coupled with Chrome's V8 JavaScript engine. The browser scores a 93/100 on the Acid3 Test.
Java support	While Android applications are written in Java, there's no Java Virtual Machine in the platform and Java byte code is not executed. Java classes get recompiled into Dalvik executable and run on Dalvik virtual machine. Dalvik is a specialized virtual machine designed specifically for Android and optimized for battery-powered mobile devices with limited memory and CPU. J2ME support can be provided via third-party-application such as the J2ME MIDP Runner.[73]
Media support	Android supports the following audio/video/still media formats: H.263, H.264 (in 3GP or MP4 container), MPEG-4 SP, AMR, AMR-WB (in 3GP container), AAC, HE-AAC (in MP4 or 3GP container), MP3, MIDI, Ogg Vorbis, WAV, JPEG, PNG, GIF, BMP.[72]
Streaming media support	RTP/RTSP streaming (3GPP PSS, ISMA), HTML progressive download (HTML5 <video> tag). Adobe Flash Streaming (RTMP) is supported through Adobe Flash Player plugin. Apple HTTP Live Streaming is supported through third party media player (Nextreaming NexPlayer). Microsoft Smooth Streaming is planned to be supported through the awaited port of Silverlight plugin to Android. Adobe Flash HTTP Dynamic Streaming is planned to be supported through an upgrade of the Flash plugin.
Additional hardware support	Android can use video/still cameras, touchscreens, GPS, accelerometers, gyroscopes, magnetometers, proximity and pressure sensors, thermometers, accelerated 2D bit blits (with hardware orientation, scaling, pixel format conversion) and accelerated 3D graphics.

Development environment	Includes a device emulator, tools for debugging, memory and performance profiling, and a plugin for the Eclipse IDE.
Market	Like many phone-based application stores, the Android Market is a catalog of applications that can be downloaded and installed to target hardware over-the-air, without the use of a PC. Originally only free applications were supported. Paid-for applications have been available on the Android Market in the United States since 19 February 2009.[74] The Android Market has been expanding rapidly. As of August 3, 2010, it had over 100,000 Android applications for download.[75] There are other markets, such as SlideME and Getjar.
Multi-touch	Android has native support for multi-touch which was initially made available in handsets such as the HTC Hero. The feature was originally disabled at the kernel level (possibly to avoid infringing Apple's patents on touch-screen technology).[76] Google has since released an update for the Nexus One and the Motorola Droid which enables multi-touch natively.[77]
Bluetooth	Support for A2DP and AVRCP were added in version 1.5;[39] sending files (OPP) and accessing the phone book (PBAP) were added in version 2.0;[45] and voice dialing and sending contacts between phones were added in version 2.2.[52]
Videocalling	The mainstream Android version doesn't support videocalling,[78] however some handsets could have a customized version of the operating system which supports it (like the Samsung i9000 Galaxy S and HTC Evo 4G).
Multitasking	Multitasking of applications is available.[79]
Voice based features	Google search through Voice is available as Search Input since initial release.[80] Also launched Voice actions supported on Android 2.2 onwards.
Tethering	Android supports tethering, which allows a phone to be used as a wireless/wired hotspot (All 2.2 Froyo phones, unofficial on phones running 1.6 or higher via applications available in the Android Market, e.g. PdaNet). To allow a laptop to share the 3G connection on an Android phone software may need to be installed on both the phone and the laptop[81]

Hardware running Android

The Android OS can be used as an operating system for cellphones, netbooks and tablet PCs, including the Dell Streak, Samsung Galaxy Tab and other devices.[82] [83]

The world's first TV running Android, called Scandinavia, has also been launched by the company People of Lava.[84]

The first commercially available phone to run the Android operating system was the HTC Dream, released on 22 October 2008.[85]

Software development

Early Android device.

The early feedback on developing applications for the Android platform was mixed.[86] Issues cited include bugs, lack of documentation, inadequate QA infrastructure, and no public issue-tracking system. (Google announced an issue tracker on 18 January 2008.)[87] In December 2007, MergeLab mobile startup founder Adam MacBeth stated, *"Functionality is not there, is poorly documented or just doesn't work... It's clearly not ready for prime time."*[88] Despite this, Android-targeted applications began to appear the week after the platform was announced. The first publicly available application was the Snake game.[89] [90] The Android Dev Phone is a SIM-unlocked and hardware-unlocked device that is designed for advanced developers. While developers can use regular consumer devices purchased at retail to test and use their applications, some developers may choose not to use a retail device, preferring an unlocked or no-contract device.

Software development kit

The Android SDK includes a comprehensive set of development tools.[91] These include a debugger, libraries, a handset emulator (based on QEMU), documentation, sample code, and tutorials. Currently supported development platforms include x86-architecture computers running Linux (any modern desktop Linux distribution), Mac OS X 10.4.9 or later, Windows XP or Vista. Requirements also include Java Development Kit, Apache Ant, and Python 2.2 or later. The officially supported integrated development environment (IDE) is Eclipse (3.2 or later) using the Android Development Tools (ADT) Plugin, though developers may use any text editor to edit Java and XML files then use command line tools to create, build and debug Android applications as well as control attached Android devices (e.g., triggering a reboot, installing software package(s) remotely).[92]

A preview release of the Android software development kit (SDK) was released on 12 November 2007. On 15 July 2008, the Android Developer Challenge Team accidentally sent an email to all entrants in the Android Developer Challenge announcing that a new release of the SDK was available in a "private" download area. The email was intended for winners of the first round of the Android Developer Challenge. The revelation that Google was supplying new SDK releases to some developers and not others (and keeping this arrangement private) led to widely reported frustration within the Android developer community at the time.[93]

On 18 August 2008 the Android 0.9 SDK beta was released. This release provided an updated and extended API, improved development tools and an updated design for the home screen. Detailed instructions for upgrading are available to those already working with an earlier release.[94] On 23 September 2008 the Android 1.0 SDK (Release 1) was released.[95] According to the release notes, it included "mainly bug fixes, although some smaller features were added". It also included several API changes from the 0.9 version.

On 9 March 2009, Google released version 1.1 for the Android dev phone. While there are a few aesthetic updates, a few crucial updates include support for "search by voice, priced applications, alarm clock fixes, sending gmail freeze fix, fixes mail notifications and refreshing intervals, and now the maps show business reviews". Another important update is that Dev phones can now access paid applications and developers can now see them on the Android Market.[96]

In the middle of May 2009, Google released version 1.5 (Cupcake) of the Android OS and SDK. This update included many new features including video recording, support for the stereo Bluetooth profile, a customizable onscreen keyboard system and voice recognition. This release also opened up the AppWidget framework to third party developers allowing anyone to create their own home screen widgets.[97]

In September 2009 the "Donut" version (1.6) was released which featured better search, battery usage indicator and VPN control applet. New platform technologies included Text to Speech engine (not available on all phones), Gestures & Accessibility framework.[98]

Android Applications are packaged in .apk format (which is simply a ZIP file, with a particular internal file layout that allows it to be run in place, without unpacking.[99]) and stored under /data/app folder on the Android OS. The user can run the command adb root to access this folder as only the root has permissions to access this folder.

Android Market

On 28 August 2008 Google announced the Android Market which was available to users on 22 October 2008. Support for paid applications was available from 13 February 2009 for US and UK developers,[100] with additional support from 29 countires on 30 September 2010.[101] The Android Market is the official download location for applications and games for Android powered devices, in mobile phone the Market application is built in and integrated with each version of the OS to allow user's quick access to a range of applications and games that will be usable on their individual device.[102] There is a huge amount of games, applications and widgets available on the Android Market with the number of applications being cited in November 2010 at 160,000.[103]

App Inventor for Android

On 12 July 2010 Google announced the availability of App Inventor for Android, a Web-based visual development environment for novice programmers, based on MIT's Open Blocks Java library and providing access to Android devices' GPS, accelerometer and orientation data, phone functions, text messaging, speech-to-text conversion, contact data, persistent storage, and Web services, initially including Amazon and Twitter.[104] "We could only have done this because Android's architecture is so open," said the project director, MIT's Hal Abelson.[105] Under development for over a year,[106] the block-editing tool has been taught to non-majors in computer science at Harvard, MIT, Wellsley, and the University of San Francisco, where Professor David Wolber developed an introductory computer science course and tutorial book for non-computer science students based on App Inventor for Android.[107] [108]

Android Developer Challenge

The Android Developer Challenge was a competition for the most innovative application for Android. Google offered prizes totaling 10 million US dollars, distributed between ADC I and ADC II. ADC I accepted submissions from 2 January to 14 April 2008. The 50 most promising entries, announced on 12 May 2008, each received a $25,000 award to fund further development.[109] [110] It ended in early September with the announcement of ten teams that received $275,000 each, and ten teams that received $100,000 each.[111] ADC II was announced on 27 May 2009.[112] The first round of the ADC II closed on 6 October 2009.[113] The first-round winners of ADC II comprising the top 200 applications were announced on 5 November 2009. Voting for the second round also opened on the same day and ended on November 25. Google announced the top winners of ADC II on November 30, with SweetDreams, What the Doodle!? and WaveSecure being nominated the overall winners of the challenge.[114] [115]

Google applications

Google has also participated in the Android Market by offering several applications for its services. These applications include Google Voice for the Google Voice service, Sky Map for watching stars, Finance for their finance service, Maps Editor for their MyMaps service, Places Directory for their Local Search, Google Goggles that searches by image, Gesture Search for using finger written letters and numbers to search the contents of the phone, Google Translate, Google Shopper, Listen for podcasts and My Tracks, a jogging application.

In August 2010, Google launched "Voice Actions for Android",[116] which allows users to search, write messages, and initiate calls by voice.

Third party applications

With the growing number of Android handsets, there has also been an increased interest by third party developers to port their applications to the Android operating system. Notable applications that have been converted to the Android operating system include Shazam, Doodle Jump, and WeatherBug.

The Android operating system has grown significantly, and a lot of the most popular internet sites and services have created native applications. These include MySpace, Facebook, and Twitter.

As of 15 July 2010, the Android Marketplace had over 70,000 applications, with over 1 billion downloads.[117] [118]

Native code

Libraries written in C and other languages can be compiled to ARM native code and installed using the Android Native Development Kit. Native classes can be called from Java code running under the Dalvik VM using the System.loadLibrary call, which is part of the standard Android Java classes.[119] [120]

Complete applications can be compiled and installed using traditional development tools.[121] The ADB debugger gives a root shell under the Android Emulator which allows native ARM code to be uploaded and executed. ARM code can be compiled using GCC on a standard PC.[121] Running native code is complicated by the fact that Android uses a non-standard C library (libc, known as Bionic). The underlying graphics device is available as a framebuffer at */dev/graphics/fb0*.[122] The graphics library that Android uses to arbitrate and control access to this device is called the Skia Graphics Library (SGL), and it has been released under an open source license.[123] Skia has backends for both win32 and Unix, allowing the development of cross-platform applications, and it is the graphics engine underlying the Google Chrome web browser.[124]

Community-based firmware

There is a community of open-source enthusiasts that build and share Android-based firmware with a number of customizations and additional features, such as FLAC lossless audio support and the ability to store downloaded applications on the microSD card.[125] This usually involves rooting the device. Rooting allows users root access to the operating system, giving more control over their environment variables. In order to use custom firmwares the devices bootloader must be unlocked. Rooting alone does not allow the flashing of custom firmware. Modified firmwares allow users of older phones to use applications available only on newer releases.[126]

Those firmware packages are updated frequently, incorporate elements of Android functionality that haven't yet been officially released within a carrier-sanctioned firmware, and tend to have fewer limitations. CyanogenMod and VillainROM are two examples of such firmware.

On 24 September 2009, Google issued a cease and desist letter[127] to the modder Cyanogen, citing issues with the re-distribution of Google's closed-source applications[128] within the custom firmware. Even though most of Android OS is open source, phones come packaged with closed-source Google applications for functionality such as the application store and GPS navigation. Google has asserted that these applications can only be provided through approved distribution channels by licensed distributors. Cyanogen has complied with Google's wishes and is continuing to distribute this mod without the proprietary software. He has provided a method to back up licensed Google applications during the mod's install process and restore them when it is complete.[129]

Marketing

Logos

The Android logo was designed with the Droid font family made by Ascender Corporation.[130]

Android Green is the color of the Android Robot that represents the Android operating system. The print color is PMS 376C and the RGB color value in hexadecimal is #A4C639, as specified by the Android Brand Guidelines.[131]

Android robot logo.

Typeface

The custom typeface of Android is called Norad, only used in the text logo.[132]

Text logo.

Market share

Research company Canalys estimates that by Q2 2009, Android had a 2.8% share of the worldwide smartphone market.[133] By the following quarter (Q3 2009), Android's market share had grown to 3.5%.[134]

In February 2010 ComScore ranked the Android platform as obtaining a 9.0% of the smartphone platform marketshare. This figure was up from an earlier estimate of 5.2% stated in November 2009.[135] In July 2010 ComScore revised Android's share for 3 months March/April/May 2010 to 13.0%, an increase of 4 percentage points, 0.2 percentage points behind Microsoft whose share had dropped 1.9%.[136]

Analytics firm Flurry estimates that 250,000 Motorola Droid phones were sold in the United States during the phone's first week in stores.[137]

In May 2010, Android's first quarter U.S. sales surpassed that of the rival iPhone platform. According to a report by the NPD group, Android achieved 28% smartphone sales in the US market, up 8% from the December quarter. In the second quarter, Apple's iOS was up by 1%, indicating that Android is taking market share mainly from RIM, and still has to compete with heavy consumer demand for new competitor offerings.[9] Furthermore, analysts point to advantages that Android has as multi-channel, multi-carrier OS, which has allowed it to duplicate the quick success of Microsoft's Windows Mobile.[138]

According to an interview with Eric Schmidt in *The Guardian*, Android is getting 160,000 new users per day (end June 2010) up from 100,000 per day in May 2010.[139]

As of 4 August 2010 Google is now activating 200,000 new phones to the Android platform per day according to Eric Schmidt.[140]

In early October 2010, Google added 20 countries to its list of approved submitters. By mid-October, purchasing apps will be available in a total of 32 countries.[141]

Market research firm Gartner reported that at the end of the Q3 2010, Android had a worldwide market share of 25.5%.[142]

For a complete list of countries that are allowed to sell apps and those able to buy them see Android Market.

Android OS usage share

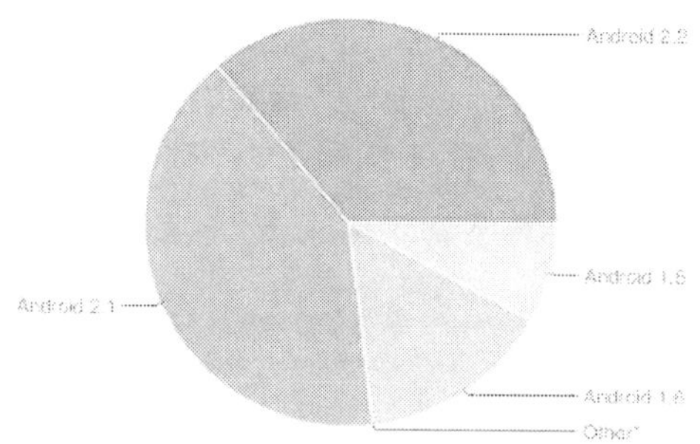

Data collected during two weeks ending on 1 November 2010
Other: 0.1% of devices running obsolete versions[143]

Platform	API Level	Distribution
Android 2.2 (Froyo)	8	36.2%
Android 2.1 (Eclair)	7	40.8%
Android 1.6 (Donut)	4	15.0%
Android 1.5 (Cupcake)	3	7.9%

Restrictions and issues

Google tracks issues and feature requests publicly at Google Code's site.[144]

Linux compatibility

- Android's kernel was derived from Linux but has been tweaked by Google outside the main Linux kernel tree.[145] Android does not have a native X Window System nor does it support the full set of standard GNU libraries, and this makes it difficult to port existing GNU/Linux applications or libraries to Android.[146] However, support for the X Window System is possible.[147]
- Google no longer maintains the code they previously contributed to the Linux kernel as part of their Android effort, effectively branching kernel code in their own tree, separating their code from Linux.[148] [149] [150] This was due to a disagreement about new features Google felt were necessary (some related to security of mobile applications) . The code which is no longer maintained was deleted in January 2010 from the Linux codebase.[151] However, Google announced in April 2010 that they will employ staff to work with the Linux kernel community.[152]

Networking issues

- Support for setting up a network proxy configuration for WiFi connections is not available.[153]
- Support for setting up a network proxy configuration for APN (i.e. GSM/EDGE) connections is not available.[154]
- Android doesn't natively support EAP extensions configuration.[155]
- Android does not support Cisco virtual private network servers requiring XAUTH extensions for IPsec (L2TP/IPsec and PPTP are supported).[156]

Issues concerning application development

- Android does not use established Java standards, i.e. Java SE and ME. This prevents compatibility among Java applications written for those platforms and those for the Android platform. Android only reuses the Java language syntax, but does not provide the full-class libraries and APIs bundled with Java SE or ME.[157] However, the Myriad Group claim that their new J2Android tool can convert Java MIDlets into Android applications.[158] [159] [160]
- Developers have reported that it is difficult to maintain applications on multiple versions of Android, owing to compatibility issues between versions 1.5 and 1.6,[161] [162] especially the different resolution ratios in use among various Android phones.[163] Such problems were pointedly brought into focus as they were encountered during the ADC2 contest.[164]
- The rapid growth in the number of Android-based phone models with differing hardware capabilities also makes it difficult to develop applications that work on all Android-based phones.[165] [166] [167] [168] As of August 2010, 64% of Android phones run the 2.x versions, and 36% still run the 1.5 and 1.6 versions[169]

Other issues

- Older versions of Android do not readily support Bluetooth file exchange,[170] although it may still be achieved with some hacking.[171] Bluetooth is supported by more recent phones.[172]
- In version 2.2 the rSAP protocol is missing which many vehicles use for handsfree.[173]
- Using the native Google Calendar functionality for Android phones, an Android device user runs into the same limitations that exist in the Calendar application. The most noticeable defect is the lack of proper time zone support: it is not possible to set the time zone for start/end times of events.[174] [175] [176] Because of this issue, some users experience difficulty while traveling with Android devices.[177]
- As of the 2.2 release, Android does not have full Unicode support.[178] Developers are reporting rendering issues, support for conjunct consonants, etc.[179]
- Android supports all the file systems supported by the linux kernel, with its own limitations. For read/write access to other popular filesystems, Tuxera launched Tuxera File System Suite, which combines NTFS, exFAT and HFS+ for Android.[180]

Claimed infringement of copyrights and patents

On the 12 August 2010, Oracle, owner of Java since it acquired Sun Microsystems in April 2009, sued Google over claimed infringement of copyrights and patents. The lawsuit claims that, "In developing Android, Google knowingly, directly and repeatedly infringed Oracle's Java-related intellectual property."[181] Oracle has named Boies, Schiller & Flexner as part of its legal team.[182]

Specifically the patent infringement claim references seven patents including United States Patent No. 5,966,702, entitled "Method And Apparatus For Preprocessing And Packaging Class Files", and United States Patent No. 6,910,205, entitled "Interpreting Functions Utilizing A Hybrid Of Virtual And Native Machine Instructions".[183] It also references United States Patent No. RE38,104, ("the '104 patent") entitled "Method And Apparatus For Resolving Data References In Generated Code" authored by James Gosling best known as the father of the Java programming language.[184]

According to Gartner analyst Ken Dulaney, Android is based on a clean room reverse-engineered version of Java, called Dalvik, which was developed without using any Sun technology or intellectual property. Oracle says Dalvik is a competitor to Java and infringes several of its patents, which are listed in the complaint, and its Java copyright.[181][185] While officially claiming that "Android is not Java", Google at the same time calls the suit an "attack on *Java* community",[186] making a distinction between "official Java" and "Java in general".

The Free Software Foundation has said that Google could have avoided this suit by building Android on top of IcedTea whose GPL license provides some protection against patents, instead of implementing it independently under the Apache License. It has also called the suit a "clear attack against someone's freedom to use, share, modify, and redistribute software".[187] However, the FSF also criticized Google, writing that "It's sad to see that Google apparently shunned those protections in order to make proprietary software development easier on Android.", and remarking that Google had not taken any clear position or action against software patents.

See also

- Android Market
- BlackBerry OS
- Dalvik virtual machine
- Chromium OS
- Google Chrome OS
- iOS
- LiMo Foundation
- Linux Phone Standards Forum
- List of Android devices
- List of Android OS-related topics
- List of Open Source Android Applications
- MeeGo Linux
- Nexus One
- Goobuntu
- Samsung's Bada OS
- Mobile World Congress
- Mobilinux
- OPhone
- Open Mobile Alliance
- Openmoko
- Palm, Inc.'s webOS
- Symbian Foundation
- Windows Mobile
- Windows Phone 7
- Google TV
- Droid (from Motorola)

References

[1] Lextrait, Vincent (January 2010). "The Programming Languages Beacon, v10.0" (http://www.lextrait.com/Vincent/implementations.html). . Retrieved 2010-01-05.

[2] "Nexus One gets an update to FRG83D" (http://www.androidcentral.com/nexus-one-gets-update-frg83d-no-its-still-not-gingerbread). Android Central. . Retrieved 2010-11-18.

[3] "Licenses" (http://source.android.com/source/licenses.html). *Android Open Source Project*. Open Handset Alliance. . Retrieved 2010-06-10.

[4] http://www.android.com/

[5] "Google Buys Android for Its Mobile Arsenal" (http://www.businessweek.com/technology/content/aug2005/tc20050817_0949_tc024.htm). Businessweek.com. 2005-08-17. . Retrieved 2010-10-29.

[6] Open Handset Alliance (2007-11-05). "Industry Leaders Announce Open Platform for Mobile Devices" (http://www.openhandsetalliance.com/press_110507.html). Press release. . Retrieved 2007-11-05.

[7] Open Handset Alliance. "Open Handset Alliance - FAQ" (http://www.openhandsetalliance.com/oha_faq.html). Press release. . Retrieved 2010-11-15.

[8] "About the Android Open Source Project" (http://source.android.com/about/index.html). . Retrieved 2010-11-15.

[9] "Android hits top spot in U.S. smartphone market" (http://news.cnet.com/8301-1035_3-20012627-94.html). 2010-08-04. . Retrieved 2010-08-04.

[10] Gabriel Madway (2010-08-04). "Google's Android leads U.S. smartphones" (http://ca.reuters.com/article/businessNews/idCATRE6734HB20100804). *Reuters*. . Retrieved 2010-08-04.

[11] "Android Most Popular Operating System in U.S. Among Recent Smartphone Buyers | Nielsen Wire" (http://blog.nielsen.com/nielsenwire/online_mobile/android-most-popular-operating-system-in-u-s-among-recent-smartphone-buyers/). Blog.nielsen.com. 2010-10-05. . Retrieved 2010-10-29.

[12] http://www.mercurynews.com/top-stories/ci_16493024?nclick_check=1

[13] "Android Dev Twitter: One Hundred Apps" (http://twitter.com/#!/AndroidDev/status/28701488389). 2010-10-26. . Retrieved 2010-10-26.

[14] David Murphy. "Extrapolating the Apple-Android Showdown: Who's Right?" (http://www.pcmag.com/article2/0,2817,2366624,00.asp). pcmag.com. . Retrieved 2010-08-24.

[15] Shankland, Stephen (12 November 2007). "Google's Android parts ways with Java industry group" (http://www.news.com/8301-13580_3-9815495-39.html). *CNET News*. .

[16] "Open Handset Alliance" (http://www.openhandsetalliance.com/). Open Handset Alliance. . Retrieved 2010-06-10.

[17] Jackson, Rob (10 December 2008). "Sony Ericsson, HTC Androids Set For Summer 2009" (http://phandroid.com/2008/12/10/sony-ericsson-htc-androids-set-for-summer-2009/). *Android Phone Fans*. . Retrieved 2009-09-03.

[18] "Android Overview" (http://www.openhandsetalliance.com/android_overview.html). Open Handset Alliance. . Retrieved 2008-09-23.

[19] "Open Core" (http://www.opencore.net/). . Retrieved 2010-06-03.

[20] Gubatron.com (23 May 2010). "How many lines of code does it take to create the Android OS?" (http://www.gubatron.com/blog/2010/05/23/how-many-lines-of-code-does-it-take-to-create-the-android-os/). . Retrieved 2010-06-03.

[21] Elgin, Ben (2005-08-17). "Google Buys Android for Its Mobile Arsenal" (http://www.businessweek.com/technology/content/aug2005/tc20050817_0949_tc024.htm). Business Week. . Retrieved 2007-11-07.

[22] Markoff, John (2007-11-04). "I, Robot: The Man Behind the Google Phone" (http://www.nytimes.com/2007/11/04/technology/04google.html?_r=2&hp=&pagewanted=all). New York Times. . Retrieved 2008-10-14.

[23] Kirsner, Scott (2007-09-02). "Introducing the Google Phone" (http://www.boston.com/business/technology/articles/2007/09/02/introducing_the_google_phone/). *The Boston Globe*. . Retrieved 2008-10-24.

[24] Nokia (23 September 2003). "T-Mobile Brings Unlimited Multiplayer Gaming to US Market with First Launch of Nokia N-Gage Game Deck" (http://www.nokia.com/A4136002?newsid=918410). Press release. . Retrieved 2009-04-05.

[25] Elgin, Ben (17 August 2005). "Google Buys Android for Its Mobile Arsenal" (http://www.businessweek.com/technology/content/aug2005/tc20050817_0949_tc024.htm). *BusinessWeek*. . Retrieved 2009-04-23.

[26] Block, Ryan (2007-08-28). "Google is working on a mobile OS, and it's due out shortly" (http://www.engadget.com/2007/08/28/google-is-working-on-a-mobile-os-and-its-due-out-shortly/). *Engadget*. . Retrieved 2007-11-06.

[27] Sharma, Amol; Delaney, Kevin J. (2007-08-02). "Google Pushes Tailored Phones To Win Lucrative Ad Market" (http://online.wsj.com/article_email/SB118602176520985718-lMyQjAxMDE3ODA2MjAwMjIxWj.html). *The Wall Street Journal*. . Retrieved 2007-11-06.

[28] "Google admits to mobile phone plan" (http://www.directtraffic.org/OnlineNews/Google_admits_to_mobile_phone_plan_18094880.html). *directtraffic.org*. Google News. 2007-03-20. . Retrieved 2007-11-06.

[29] McKay, Martha (21 December 2006). "Can iPhone become your phone?; Linksys introduces versatile line for cordless service". *The Record*: p. L9. "And don't hold your breath, but the same cell phone-obsessed tech watchers say it won't be long before Google jumps headfirst into the phone biz. Phone, anyone?"

[30] Ackerman, Elise (2007-08-30). "Blogosphere Aflutter With Linux-Based phone Rumors" (http://www.linuxinsider.com/rsstory/59115.html). *Linux Insider*. . Retrieved 2007-11-07.

[31] Claburn, Thomas (2007-09-19). "Google's Secret Patent Portfolio Predicts gPhone" (http://www.informationweek.com/news/showArticle.jhtml?articleID=201807587&cid=nl_IWK_daily). *InformationWeek*. . Retrieved 2007-11-06.
[32] Pearce, James Quintana (2007-09-20). "Google's Strong Mobile-Related Patent Portfolio" (http://www.moconews.net/entry/419-googles-strong-mobile-related-patent-portfolio/). *mocoNews.net*. . Retrieved 2007-11-07.
[33] Martinez, Jennifer (2008-12-10). "CORRECTED — UPDATE 2-More mobile phone makers back Google's Android" (http://www.reuters.com/article/newsOne/idUSN0928595620081210). *Reuters* (Thomson Reuters). . Retrieved 2008-12-13.
[34] Kharif, Olga (2008-12-09). "Google's Android Gains More Powerful Followers" (http://www.businessweek.com/the_thread/techbeat/archives/2008/12/googles_android_2.html). *BusinessWeek*. McGraw-Hill. . Retrieved 2008-12-13.
[35] Boulton, Clint (21 October 2008). "Google Open-Sources Android on Eve of G1 Launch" (http://www.eweek.com/c/a/Mobile-and-Wireless/Google-Open-Sources-Android-on-Eve-of-G1-Launch/). *eWeek*. . Retrieved 2009-09-03.
[36] Bort, Dave (21 October 2008). "Android is now available as open source" (http://source.android.com/posts/opensource). *Android Open Source Project*. . Retrieved 2009-09-03.. Mirror link (https://sites.google.com/a/android.com/opensource/posts/opensource).
[37] Ducrohet, Xavier (27 April 2009). "Android 1.5 is here!" (http://android-developers.blogspot.com/2009/04/android-15-is-here.html). *Android Developers Blog*. . Retrieved 2009-09-03.
[38] Rob, Jackson (30 April 2009). "CONFIRMED: Official Cupcake Update Underway for T-Mobile G1 USA & UK!" (http://phandroid.com/2009/04/30/official-cupcake-update-underway-for-t-mobile-g1-usa/). *Android Phone Fans*. . Retrieved 2009-09-03.
[39] "Android 1.5 Platform Highlights" (http://developer.android.com/sdk/android-1.5-highlights.html). *Android Developers*. April 2009. . Retrieved 2009-09-03.
[40] "Android 1.6 Platform Highlights" (http://developer.android.com/sdk/android-1.6-highlights.html). *Android Developers*. September 2009. . Retrieved 2009-10-01.
[41] Ducrohet, Xavier (15 September 2009). "Android 1.6 SDK is here" (http://android-developers.blogspot.com/2009/09/android-16-sdk-is-here.html). *Android Developers Blog*. . Retrieved 2009-10-01.
[42] Ryan, Paul (1 October 2009). "Google releases Android 1.6; Palm unleashes WebOS 1.2" (http://arstechnica.com/gadgets/news/2009/10/google-releases-android-16-palm-releases-webos-12.ars). *Ars Technica*. . Retrieved 2009-10-01.
[43] "Android 2.1 / Eclair on Google Nexus One" (http://www.google.com/phone/static/en_US-nexusone_tech_specs.html). *Android Developers*. . Retrieved 2010-01-05. (Eclair)
[44] "Android 2.0, Release 1" (http://developer.android.com/sdk/android-2.0.html). *Android Developers*. . Retrieved 2009-10-27.
[45] "Android 2.0 Platform Highlights" (http://developer.android.com/sdk/android-2.0-highlights.html). *Android Developers*. . Retrieved 2009-10-27.
[46] "Android 2.0 API Changes Summary" (http://developer.android.com/sdk/android-2.0.html#api-changes). . Retrieved 2010-03-06.
[47] "Android 2.0.1, Release 1" (http://developer.android.com/sdk/android-2.0.1.html). *Android Developers*. . Retrieved 2010-01-17.
[48] "Android 2.1, Release 1" (http://developer.android.com/sdk/android-2.1.html). *Android Developers*. . Retrieved 2010-01-17.
[49] Savov, Vladislav (16 January 2010). "Next Android version will be called Froyo, says Erick Tseng" (http://www.engadget.com/2010/01/16/next-android-version-will-be-called-froyo-says-erick-tseng/). *Engadget.com*. . Retrieved 2010-01-16.
[50] Swetland, Brian (7 February 2010). "Some clarification on "the Android Kernel"" (http://lwn.net/Articles/373374/). *lwn.net*. . Retrieved 2010-02-21.
[51] Ducrohet, Xavier (20 May 2010). "Android 2.2 and developers goodies." (http://android-developers.blogspot.com/2010/05/android-22-and-developers-goodies.html). *Android Developers Blog*. Google. . Retrieved 2010-05-20.
[52] "Android 2.2 Platform Highlights" (http://developer.android.com/sdk/android-2.2-highlights.html). *Android Developers*. 20 May 2010. . Retrieved 2010-05-23.
[53] "Unofficially Confirmed Froyo Features, Post-Day-1 Of Google I/O" (http://www.androidpolice.com/2010/05/20/exclusive-unofficially-confirmed-froyo-features-post-day-1-of-google-io-google-io-blitz-coverage-day-1/). *Android Police*. . Retrieved 2010-05-20.
[54] "Nexus One Is Running Android 2.2 Froyo. How Fast Is It Compared To 2.1? Oh, Only About 450% Faster" (http://www.androidpolice.com/2010/05/11/exclusive-androidpolice-coms-nexus-one-is-running-android-2-2-froyo-how-fast-is-it-compared-to-2-1-oh-only-about-450-faster/). *androidpolice*. . Retrieved 2010-05-13.
[55] "Browser support for file upload field is coming in Froyo" (http://code.google.com/p/android/issues/detail?id=2519#c112). *Google Code*. . Retrieved 2010-05-13.
[56] "Android support for memory card app storage is finally "coming soon"" (http://code.google.com/p/android/issues/detail?id=1151#c535). *Google Code*. . Retrieved 2010-05-01.
[57] Stone, Brad (2010-04-27). "Google's Andy Rubin on Everything Android" (http://bits.blogs.nytimes.com/2010/04/27/googles-andy-rubin-on-everything-android/). *NY Times*. . Retrieved 2010-05-20.
[58] "Gingerbread to Be Android 2.3, Statue Arrives at Google Campus - Softpedia" (http://news.softpedia.com/news/Gingerbread-to-Be-Android-2-3-Statue-Arrives-at-Google-Campus-162540.shtml). News.softpedia.com. 2010-10-23. . Retrieved 2010-10-29.
[59] "When will other Google products support WebM and VP8?" (http://www.webmproject.org/about/faq/#webm_video_file_format). *The WebM Project*. . Retrieved 2010-05-20.

[60] "Google's CEO on the Future of Mobile and More [LIVE (http://mashable.com/2010/11/15/google-ceo-web-2/)"]. *The WebM Project*. . Retrieved 2010-05-20.

[61] "Issue 3190: Improve copy-paste in Browser/WebView" (http://code.google.com/p/android/issues/detail?id=3190#c81). *Google Code*. . Retrieved 2010-05-20.

[62] Segan, Sascha (2010-10-08). "Exclusive Q&A: Google's Andy Rubin Talks Android | News & Opinion" (http://www.pcmag.com/article2/0,2817,2370464,00.asp). PCMag.com. . Retrieved 2010-10-29.

[63] "Android Music Store to Take on Apple iTunes" (http://www.pcworld.com/article/196871/android_music_store_to_take_on_apple_itunes.html). .

[64] "Android Team "Laser Focused" On The User Experience For Next Release" (http://techcrunch.com/2010/06/16/android-team-laser-focused-on-the-user-experience-for-next-release/). .

[65] "Android 3.0 Gingerbread details: 1280×760 resolution, 1Ghz minimum specs, Q4 release" (http://www.unwiredview.com/2010/06/30/android-3-0-gingerbread-details-1280x760-resolution-1ghz-minimum-specs-mid-oct-release/). .

[66] commit merging mksh into the tree (http://android.git.kernel.org/?p=platform/system/core.git;a=commit;h=ba2627c6cdb3aaa40aebd362170c382b55b7b511) – /system/bin/sh is the Unix /bin/sh equivalent for Android

[67] Beavis, Gareth (19 August 2010). "Revealed: Android Honeycomb next up from Google" (http://www.techradar.com/news/phone-and-communications/mobile-phones/revealed-android-honeycomb-next-up-from-google-711132). *TechRadar*. .

[68] "Google Android 2.3 Gingerbread Official: 3.0 is Honeycomb" (http://www.onlinesocialmedia.net/20101023/google-android-2-3-gingerbread-official-3-0-is-honeycomb/). Online Social Media. 2010-10-23. . Retrieved 2010-10-29.

[69] Woyke, Elizabeth (15 October 2010). "Next Android Release To Be Called 'Ice Cream'" (http://blogs.forbes.com/elizabethwoyke/2010/10/15/next-android-release-to-be-called-ice-cream/?boxes=techchanneltopstories). *Forbes*. . Retrieved 2010-11-03.

[70] "What is Android?" (http://developer.android.com/guide/basics/what-is-android.html). *Android Developers*. 21 July 2009. . Retrieved 2009-09-03.

[71] Topolsky, Joshua (2007-11-12). "Google's Android OS early look SDK now available" (http://www.engadget.com/2007/11/12/googles-android-os-early-look-sdk-now-available/). *Engadget*. . Retrieved 2007-11-12.

[72] "Android Supported Media Formats" (http://developer.android.com/guide/appendix/media-formats.html). *Android Developers*. . Retrieved 2009-05-01.

[73] Android J2ME MIDP RUNNER homepage (http://www.netmite.com/android/)

[74] Patel, Nilay (19 February 2009). "Paid apps appear in Android Market" (http://www.engadget.com/2009/02/19/paid-apps-appear-in-android-market/). *Engadget*. . Retrieved 2009-04-06.

[75] Laura June (July 15, 2010). "Android Market now has 100,000 apps, passes 1 billion download mark (update: Google says 70K)" (http://www.engadget.com/2010/07/15/android-market-now-has-100-000-apps-passes-1-billion-download-m/). engadget.com. . Retrieved 2010-08-24.

[76] Musil, Steven (11 February 2009). "Report: Apple nixed Android's multitouch" (http://news.cnet.com/8301-13579_3-10161312-37.html). *CNET News*. . Retrieved 2009-09-03.

[77] Ziegler, Chris (2 February 2010). "Nexus One gets a software update, enables multitouch" (http://www.engadget.com/2010/02/02/nexus-one-gets-a-software-update-enables-multitouch/). *Engadget*. . Retrieved 2010-02-02.

[78] Issue 2664 (http://code.google.com/p/android/issues/detail?id=2664), first reported 14 May 2009, unreviewed as of October 2010.

[79] Bray, Tim (28 April 2010). "Multitasking the Android Way" (http://android-developers.blogspot.com/2010/04/multitasking-android-way.html). *Android Developers*. . Retrieved 2010-11-03.

[80] "Speech Input for Google Search" (http://developer.android.com/resources/articles/speech-input.html). *Android Developers*. . Retrieved 2010-11-03.

[81] JR Raphael (May 6, 2010). "Use Your Android Phone as a Wireless Modem" (http://www.pcworld.com/article/190265/use_your_android_phone_as_a_wireless_modem.html). PCWorld. . Retrieved 2010-11-03.

[82] "Basics of Android" (http://www.beinno.com/2010/09/basics-about-android.html). beinno.com. . Retrieved 2010-11-03.

[83] "Unboxing Acer Aspire One Android netbook" (http://www.youtube.com/watch?v=U1e0q2DYAd8). YouTube. 2009-11-25. . Retrieved 2010-10-29.

[84] "People of Lava - Scandinavia, the World's first Android TV" (http://www.peopleoflava.com/television/scandinavia/). People of Lava. . Retrieved 2010-11-03.

[85] "T-Mobile Unveils the T-Mobile G1 - the First Phone Powered by Android" (http://www.htc.com/www/press.aspx?id=66338&lang=1033). HTC. . Retrieved 2009-05-19. AT&T's first device to run the Android OS was the Motorola Backflip.

[86] Paul, Ryan (2007-12-19). "Developing apps for Google Android: it's a mixed bag" (http://arstechnica.com/news.ars/post/20071219-google-android-plagued-by-dysfunctional-development-process.html). *Ars Technica*. . Retrieved 2007-12-19.

[87] Morrill, Dan (18 January 2008). "You can't rush perfection, but now you can file bugs against it" (http://android-developers.blogspot.com/2008/01/you-cant-rush-perfection-but-now-you.html). *Android Developers Blog*. . Retrieved 2009-09-03.

[88] Morrison, Scott (2007-12-19). "Glitches Bug Google's Android Software" (http://online.wsj.com/article_email/SB119800856883537515-lMyQjAxMDE3OTE4ODAxMDg4Wj.html). *The Wall Street Journal*. . Retrieved 2007-12-19.

[89] "Snake" (http://www.android-freeware.org/download/snake). *Android Freeware Directory*. . Retrieved 2008-01-26.

[90] "First Android Application — Snake" (http://www.mobiles2day.com/2007/11/14/first-android-application-snake/). *Mobiles2day*. 2007-11-14. . Retrieved 2008-01-07.

[91] "Tools Overview" (http://developer.android.com/guide/developing/tools/index.html). *Android Developers*. 21 July 2009. .

[92] Westfall, Jon (25 August 2009). "Backup & Restore Android Apps Using ADB" (http://jonwestfall.com/2009/08/backup-restore-android-apps-using-adb/). *JonWestfall.com*. . Retrieved 2009-12-07.

[93] Metz, Cade (14 July 2008). "Google plays Hide and Seek with Android SDK" (http://www.theregister.co.uk/2008/07/14/android_developer_unrest/). *The Register*. . Retrieved 2008-10-23.

[94] "Android — An Open Handset Alliance Project: Upgrading the SDK" (http://code.google.com/android/intro/upgrading.html). . Retrieved 2008-10-24.

[95] "Other SDK Releases" (http://developer.android.com/sdk/older_releases.html). *Android Developers*. . Retrieved 2009-09-02.

[96] Jackson, Rob (9 March 2009). "Android Dev Phone Update: Version 1.1!" (http://phandroid.com/2009/03/09/android-dev-phone-update-version-11/). *Android Phone Fans*. . Retrieved 2009-09-03.

[97] "Android 1.5 Version Notes" (http://developer.android.com/sdk/android-1.5.html#features). *Android Developers*. April 2009. . Retrieved 2009-09-03.

[98] "Android 1.6 Version Notes" (http://developer.android.com/sdk/android-1.6-highlights.html). *Android Developers*. September 2009. . Retrieved 2009-09-03.

[99] "Android explained" (http://heartofthings.blogspot.com/2010/11/android.html). .

[100] "Android Market Update Support" (http://android-developers.blogspot.com/2009/02/android-market-update-support-for.html). .

[101] "More Countries More Sellers More Buyers" (http://android-developers.blogspot.com/2010/09/more-countries-more-sellers-more-buyers.html). .

[102] "What Is Android?" (http://recombu.com/news/what-is-android-and-what-is-an-android-phone_M12615.html). .

[103] "Android App Stats" (http://www.androlib.com/appstats.aspx). .

[104] Claburn, Thomas (2010-07-12). "Google App Inventor Simplifies Android Programming" (http://www.informationweek.com/news/smb/mobile/showArticle.jhtml?articleID=225702880&subSection=News). *Information Week*. . Retrieved 2010-07-12.

[105] Lohr, Steve (2010-07-11). "Google's Do-It-Yourself App Creation Software" (http://www.nytimes.com/2010/07/12/technology/12google.html?src=busln). *New York Times*. . Retrieved 2010-07-12.

[106] Abelson, Hal (2009-07-31). "App Inventor for Android" (http://googleresearch.blogspot.com/2009/07/app-inventor-for-android.html). *Google Research Blog*. . Retrieved 2010-07-12.

[107] Kim, Ryan (2009-12-11). "Google brings app-making to the masses" (http://articles.sfgate.com/2009-12-11/business/17220628_1_computer-science-smart-phone-android). *San Francisco Chronicle*. . Retrieved 2010-07-12.

[108] Wolber, David. "AppInventor.org" (http://www.appinventor.org/). . Retrieved 2010-07-12.

[109] Chen, Jason (12 May 2008). "The Top 50 Applications" (http://android-developers.blogspot.com/2008/05/top-50-applications.html). *Android Developers Blog*. . Retrieved 2009-09-04.

[110] Brown, Eric (13 May 2008). "Android Developer Challenge announces first-round winners" (http://www.linuxdevices.com/news/NS3168326017.html). *Linux for Devices*. .

[111] "ADC I Top 50 Gallery" (http://code.google.com/android/adc/adc_gallery/). *Android Developer Challenge*. . Retrieved 2009-05-19.

[112] "Android Developer Challenge" (http://code.google.com/android/adc/). *Google Code*. . Retrieved 2008-01-11.

[113] Chu, Eric (6 October 2009). "ADC 2 Round 1 Scoring Complete" (http://android-developers.blogspot.com/2009/10/adc-2-round-1-scoring-complete.html). *Android Developers Blog*. . Retrieved 2009-11-03.

[114] "ADC 2 Overall Winners" (http://code.google.com/android/adc/gallery_winners.html). *Android Developer Challenge*. Google. . Retrieved 2010-12-05.

[115] Kharif, Olga (30 November 2009). "Android Developer Challenge 2 Winners Announced" (http://www.businessweek.com/the_thread/techbeat/archives/2009/11/android_develop_2.html). *BusinessWeek*. . Retrieved 2010-12-05.

[116] Voice Actions for Android (http://www.google.com/mobile/voice-actions/index.html)

[117] "Android Market Has 100,000 Apps & Passes 1 Billion Downloads" (http://www.engadget.com/2010/07/15/android-market-now-has-100-000-apps-passes-1-billion-download-m/). 2010-07-15. . Retrieved 2010-07-15.

[118] "Android Market Htis 1 Billion Downloads & 100,000 apps" (http://www.fonehome.co.uk/2010/07/16/android-market-hits-1-billion-downloads-100000-apps/). 2010-07-15. . Retrieved 2010-07-15.

[119] Srinivas, Davanum (2007-12-09). "Android — Invoke JNI based methods (Bridging C/C++ and Java)" (http://davanum.wordpress.com/2007/12/09/android-invoke-jni-based-methods-bridging-cc-and-java/). . Retrieved 2008-12-13.

[120] "java.lang.System" (http://developer.android.com/reference/java/lang/System.html). *Android Developers*. . Retrieved 2009-09-03.

[121] Leslie, Ben (13 November 2007). "Native C application for Android" (http://benno.id.au/blog/2007/11/13/android-native-apps). *Benno's blog*. . Retrieved 2009-09-04.

[122] Cooksey, Tom (2007-11-07). "Native C *GRAPHICAL* applications now working on Android emulator" (http://groups.google.com/group/android-developers/msg/ace258af92fff692?dmode=source&pli=1). *android-developers mailing list*. . Retrieved 2008-12-13.

[123] "Skia source" (http://src.chromium.org/viewvc/chrome/trunk/src/skia/). .

[124] Toker, Alp (2008-09-06). "Skia graphics library in Chrome: First impressions" (http://www.atoker.com/blog/2008/09/06/skia-graphics-library-in-chrome-first-impressions/). . Retrieved 2008-12-13.

[125] "Dream android development" (http://forum.xda-developers.com/forumdisplay.php?f=448). *xda-developers forum*. . Retrieved 2009-09-11.

[126] "Android 2.1 from Motorola Droid Ported to G1" (http://voltmobileandtech.com/blog/). *Volt Mobile*. March 10, 2010. .

[127] Wimberly, Taylor (24 September 2009). "CyanogenMod in trouble?" (http://androidandme.com/2009/09/hacks/cyanogenmod-in-trouble/). *Android and me*. . Retrieved 2009-09-26.

[128] Morrill, Dan (25 September 2009). "A Note on Google Apps for Android" (http://android-developers.blogspot.com/2009/09/note-on-google-apps-for-android.html). *Android Developers Blog*. . Retrieved 2009-09-26.

[129] "The current state..." (http://www.cyanogenmod.com/home/the-current-state). *CyanogenMod Android Rom*. 27 September 2009. . Retrieved 2009-09-27.

[130] Woyke, Elizabeth (26 September 2008). "Android's Very Own Font" (http://www.forbes.com/2008/09/25/font-android-g1-tech-wire-cx_ew_0926font.html). *Forbes*. .

[131] "Brand Guidelines" (http://www.android.com/branding.html). *Android*. 23 March 2009. . Retrieved 2009-10-30.

[132] "Android Brand Guidelines" (http://www.android.com/branding.html). *Android*. 23 March 2009. . Retrieved 2010-04-10.

[133] "Canalys: iPhone outsold all Windows Mobile phones in Q2 2009" (http://www.appleinsider.com/articles/09/08/21/canalys_iphone_outsold_all_windows_mobile_phones_in_q2_2009.html). *AppleInsider*. 21 August 2009. . Retrieved 2009-09-21.

[134] "Canalys Q3 2009: iPhone, RIM taking over smartphone market" (http://www.appleinsider.com/articles/09/11/03/canalys_q3_2009_iphone_rim_taking_over_smartphone_market.html). *AppleInsider*. 3 November 2009. . Retrieved 2009-11-03.

[135] "Top smarthphone platforms, 3 mos. ending 2/10" (http://www.mycomscore.net/Press_Events/Press_Releases/2010/4/comScore_Reports_February_2010_U.S._Mobile_Subscriber_Market_Share). *Bloomberg Television*. comScore MobiLens. April 2010. . Retrieved 2010-04-19. "RIM, 42.1%; Apple, 25.4%; Microsoft, 15.1%; Google (Android), 9.0%; Palm, 5.4%; others, 3.0%"

[136] "comScore Reports May 2010 U.S. Mobile Subscriber Market Share - comScore, Inc" (http://www.comscore.com/Press_Events/Press_Releases/2010/7/comScore_Reports_May_2010_U.S._Mobile_Subscriber_Market_Share). Comscore.com. 2010-07-08. . Retrieved 2010-10-29.

[137] "Droid Sales and the Android Explosion" (http://www.pcworld.com/article/182310/droid_sales_and_the_android_explosion.html). *PC World*. 17 November 2009. .

[138] Greg Sandoval (2010-08-02). "More signs iPhone under Android attack" (http://news.cnet.com/8301-13579_3-20012418-37.html). . Retrieved 2010-08-04.

[139] Arthur, Charles (2010-06-25). "Eric Schmidt's dog whistle to mobile developers: abandon Windows Phone" (http://www.guardian.co.uk/technology/2010/jun/25/android-schmidt-mobile-platform). London: The Guardian. .

[140] Mark, Lightell (2010-09-01). "Google responds to Steve Jobs' activation counting accusations" (http://tech.fortune.cnn.com/2010/09/01/steve-jobs-hits-google-with-number-counting-accusations/). California?: CNN Money. .

[141] "Google expands Android's reach, accepting paid apps from 20 more countries, selling to 18 more" (http://www.engadget.com/2010/10/01/google-expands-androidss-reach-accepting-paid-apps-from-20-mor/). Engadget. 2010-10-01. . Retrieved 2010-10-29.

[142] http://www.gartner.com/it/page.jsp?id=1466313

[143] "Platform Versions" (http://developer.android.com/resources/dashboard/platform-versions.html). *Android Developers*. . Retrieved 2010-10-05.

[144] "Issues - android" (http://code.google.com/p/android/issues/list). *Google Code*. . Retrieved 2010-01-15.

[145] *Androidology - Part 1 of 3 - Architecture Overview* (http://www.youtube.com/watch?v=QBGfUs9mQYY). [Video]. YouTube. 2008-09-06. . Retrieved 2007-11-07.

[146] Paul, Ryan (23 February 2009). "Dream(sheep++): A developer's introduction to Google Android" (http://arstechnica.com/open-source/reviews/2009/02/an-introduction-to-google-android-for-developers.ars). *Ars Technica*. . Retrieved 2009-03-07. "In fact, during a presentation at the Google IO conference, Google engineer Patrick Brady stated unambiguously that Android is not Linux. (...) The problem with Google's approach is that it makes Android an island. The highly insular nature of the platform prevents Android users and developers from taking advantage of the rich ecosystem of existing third-party Linux applications. Android doesn't officially support native C programs at all, so it won't be possible to port your favorite GTK+ or Qt applications to Android"

[147] "Re:Gnome, KDE, IceWM or LXDE Desktop on your Android! - AndroidFanatic Community Forums" (http://www.androidfanatic.com/community-forums.html? func=view&catid=9&id=1615). Androidfanatic.com. . Retrieved 2010-10-29.

[148] Greg Kroah-Hartman (2010-02-02). "Android and the Linux kernel community" (http://www.kroah.com/log/linux/android-kernel-problems.html). . Retrieved 2010-02-03. "*This means that any drivers written for Android hardware platforms, cannot get merged into the main kernel tree because they have dependencies on code that only lives in Google's kernel tree, causing it to fail to build in the kernel.org tree. Because of this, Google has now prevented a large chunk of hardware drivers and platform code from ever getting merged into the main kernel tree. Effectively creating a kernel branch that a number of different vendors are now relying on.(...) But now they are stuck. Companies with Android-specific platform and drivers cannot contribute upstream, which causes these companies a much larger maintenance and development cycle.*"

[149] "Linux developer explains Android kernel code removal" (http://news.zdnet.com/2100-9595_22-389733.html). ZDNet. 2010-02-02. . Retrieved 2010-02-03.

[150] "What is Android?" (http://developer.android.com/guide/basics/what-is-android.html). *Android Developers*. . Retrieved 2010-01-08.

[151] "Android versus Linux?" (http://www.h-online.com/open/features/Android-versus-Linux-924563.html). *www.h-online.com*. 9 February 2010. . Retrieved 2010-02-28.

[152] "DiBona: Google will hire two Android coders to work with kernel.org" (http://blogs.zdnet.com/open-source/?p=6274). *www.zdnet.com*. 15 April 2010. . Retrieved 2010-04-29.

[153] Issue 1273 (http://code.google.com/p/android/issues/detail?id=1273&colspec=ID Type Status Owner Summary Stars), first reported 12 November 2008, unreviewed as of June 2010.

[154] (http://code.google.com/p/android/issues/detail?id=9126), listed as released but not resolved

[155] Issue 1386 (http://code.google.com/p/android/issues/detail?id=1386&colspec=ID Type Status Owner Summary Stars), first reported 28 November 2008, unreviewed as of June 2010.

[156] Issue 3902 (http://code.google.com/p/android/issues/detail?id=3902&colspec=ID Type Status Owner Summary Stars), first reported 15 September 2009, unreviewed as of August 2010.

[157] van Gurp, Jilles (13 November 2007). "Google Android: Initial Impressions and Criticism" (http://www.javalobby.org/nl/archive/jlnews_20071113o.html). *Javalobby*. . Retrieved 2009-03-07. *"Frankly, I don't understand why Google intends to ignore the vast amount of existing implementation out there. It seems like a bad case of "not invented here" to me. Ultimately, this will slow adoption. There are already too many Java platforms for the mobile world and this is yet another one"*

[158] "Myriad's New J2Android Converter Fuels Android Applications Gold Rush" (http://www.myriadgroup.com/Media-Centre/News/Myriad-New-J2Android-Converter-Fuels-Android-Applications-Gold-Rush.aspx). 19 March 2010. .

[159] "J2Android hopes you don't know that Android is Java-based" (http://www.javaworld.com/community/node/4170). 23 March 2010. . *"On the other hand, you might think this is kind of a scam aimed at developers who don't really understand the nature of the platform they're targeting. My biggest complaint is that you'd think that Mikael Ricknäs, the IDG News Service reporter who wrote the first story linked to above (who toils for the same company that publishes JavaWorld), would have at least mentioned the relationship between Java and Android to make the oddness of this announcement clear."*

[160] "Myriad CTO: J2Android moves MIDlets to "beautiful" Android framework" (http://www.javaworld.com/community/?q=node/4210). 31 March 2010. . *"We will have to wait and see exactly how much pickup J2Android actually sees. The tool isn't actually available on the open market just yet; while Schillings spoke optimistically about "converting 1,000 MIDlets in an afternoon," at the moment they're working with a few providers to transform their back catalogs. So those of you out there hoping to avoid learning how to write Android code may have to wait a while."*

[161] "HTC Developer center: Android Dev Phone 1" (http://developer.htc.com/adp.html). HTC Corporation. . Retrieved 2010-01-15. *"For development, you should select the lowest possible Android platform version that meets the needs of your applications. For example, if you are working in the Android 1.1 SDK and your application is using APIs introduced in Android 1.1, then you should download the Android 1.1 system image. If you are using the Android 1.1 SDK but your application does not use Android 1.1 APIs, then using Android 1.0 image is sufficient. For testing, consider downloading all platform versions with which your application is compatible, then running your applications on those platform versions to ensure that they work as designed."*

[162] "Android's Weakest Link" (http://blogs.zdnet.com/Greenfield/?p=481). ZDNet. 2009-10-11. . Retrieved 2010-01-15.

[163] "Complications looming for Android developers" (http://androidandme.com/2009/11/news/complications-looming-for-android-developers/). androidandme.com. 2009-11-06. . Retrieved 2010-01-15.

[164] "A Chink In Android's Armor" (http://www.techcrunch.com/2009/10/11/a-chink-in-androids-armor/). TechCrunch. 2009-10-11. . Retrieved 2009-10-11. *"And now they're faced with a landslide of new handsets, some running v.1.6 and some courageous souls even running android v.2.0. All those manufacturers/carriers are racing to release their phones by the 2009 holiday season, and want to ensure the hot applications will work on their phones. And here's the problem – in almost every case, we hear, there are bugs and more serious problems with the apps.[...]First of all, the compatibility between versions issue may be overblown. The reported problems have been limited to an Android developer contest[...]We haven't heard of any major app developers complaining of backwards or forward compatibility problems. Also, I've now upgraded my phone from 1.5 to 1.6, and every application continues to work fine."*

[165] "Android's Rapid Growth Has Some Developers Worried" (http://www.wired.com/gadgetlab/2009/11/android-fragmentation/). Wired News. 2009-11-16. . Retrieved 2010-02-26. *"Fagan's concerns about the fragmentation of Android is being echoed by other developers, says Sean Galligan, vice president of business development at Flurry, an mobile app analytics company(...)"You may build an app that works perfectly with all three firmwares, but then when you run it on carriers' ROMs it completely blows up," says Fagan. "So we find ourselves having to create apps that are compatible with multiple firmwares, multiple ROMs and multiple devices with different hardware."*

[166] "Android just reproducing Java ME's problems, now" (http://www.javaworld.com/community/node/3704). JavaWorld. 2009-11-17. . Retrieved 2010-02-26.

[167] "Android's Spread Could Become a Problem" (http://www.businessweek.com/technology/content/oct2009/tc20091015_626136.htm). BusinessWeek. 2009-10-15. . Retrieved 2010-02-28.

[168] "Google Android's self-destruction derby begins" (http://infoworld.com/d/mobilize/google-androids-self-destruction-derby-begins-863). InfoWorld. 2010-02-22. . Retrieved 2010-02-28.

[169] "Platform Versions" (http://developer.android.com/resources/dashboard/platform-versions.html). developer.android.com. 2010-05-03. . Retrieved 2010-08-07.

[170] "Issue 719: enhanced low-level Bluetooth support" (http://code.google.com/p/android/issues/detail?id=719). *Google Code*. . Retrieved 2010-11-03.

[171] Yueh, Erin (February 1, 2010). "New Bluetooth Object Push Profile in Android 2.0 (Eclair)" (http://i-miss-erin.blogspot.com/2010_02_01_archive.html). . Retrieved 2010-11-03.

[172] "Transferring files to the HTC Desire and HTC Legend via Bluetooth" (http://blog.brightpointuk.co.uk/transferring-files-htc-desire-and-htc-legend-bluetooth). . Retrieved 2010-11-03.

[173] "Issue 4402: rSAP / Sim access bluetooth profile" (http://code.google.com/p/android/issues/detail?id=4402). Google.com. . Retrieved 2010-11-03.
[174] "Google Calendar needs per event timezone support" (http://getsatisfaction.com/google/topics/google_calendar_needs_per_event_timezone_support). Get Satisfaction Inc. . Retrieved 2010-11-03.
[175] "FixIt!: Google Calendar and time zones" (http://www.androidguys.com/2010/01/21/fixit-google-calendar-time-zones/). AndroidGuys. January 21, 2010. . Retrieved 2010-11-03.
[176] "Help File: Google Calendar's time-zone weakness" (http://www.washingtonpost.com/wp-dyn/content/article/2010/04/03/AR2010040304956.html). The Washington Post. April 4, 2010. . Retrieved 2010-11-03.
[177] "Issue 5892: Calendar: Don't change times when moving across time zones" (http://code.google.com/p/android/issues/detail?id=5892). google.com. . Retrieved 2010-11-03.
[178] "Issue 4230: Armenian character support" (http://code.google.com/p/android/issues/detail?id=4230). google.com. . Retrieved 2010-11-03.
[179] "Issue 5925: Support full Unicode for all languages and scripts" (http://code.google.com/p/android/issues/detail?id=5925). google.com. . Retrieved 2010-11-03.
[180] "Tuxera Launches Tuxera File System Suite, First to Combine NTFS, exFAT and HFS+ for Android" (http://www.marketwire.com/press-release/Tuxera-Launches-Tuxera-File-System-Suite-First-Combine-NTFS-exFAT-HFS-Android-1342807.htm). marketwire. October 28, 2010. . Retrieved 2010-11-03.
[181] James Niccolai (2010-08-12). "Oracle sues Google over Java use in Android" (http://www.computerworld.com/s/article/9180678/Update_Oracle_sues_Google_over_Java_use_in_Android?taxonomyId=13). . Retrieved 2010-08-20.
[182] Mark Hachman. "Oracle Sues Google Over Android Java Use" (http://www.pcmag.com/article2/0,2817,2367761,00.asp). pcmag.com. . Retrieved 2010-08-24.
[183] "Oracle's complaint against Google for Java patent infringement" (http://www.scribd.com/doc/35811761/Oracle-s-complaint-against-Google-for-Java-patent-infringement). scribd.com. . Retrieved 2010-08-13.
[184] Ed Burnette (August 12, 2010). "Oracle uses James Gosling patent to attack Google and Android developers" (http://www.zdnet.com/blog/burnette/oracle-uses-james-gosling-patent-to-attack-google-and-android-developers/2035?tag=mantle_skin;content). ZD Net. . Retrieved 2010-11-03.
[185] "Android and Java comparison" (http://weblogs.java.net/blog/opinali/archive/2010/08/17/android-java). Java.net. . Retrieved 2010-11-03.
[186] Cade Metz (August 16, 2010). "Google dubs Oracle suit 'attack on Java community'" (http://www.theregister.co.uk/2010/08/16/google_oracle_android_lawsuit/). The Register. . Retrieved 2010-11-03.
[187] Brett Smith (September 8, 2010). "FSF responds to Oracle v. Google and the threat of software patents" (http://www.fsf.org/news/oracle-v-google/?searchterm=Oracle). Free Software Foundation. . Retrieved 2010-11-03.

Bibliography

- Ed, Burnette (November 10, 2009). *Hello, Android: Introducing Google's Mobile Development Platform* (http://pragprog.com/titles/eband2/hello-android) (2nd ed.). Pragmatic Bookshelf. ISBN 1934356492.
- Rogers, Rick; Lombardo, John; Mednieks, Zigurd; Meike, Blake (May 1, 2009). *Android Application Development: Programming with the Google SDK* (http://oreilly.com/catalog/9780596521509) (1st ed.). O'Reilly Media. ISBN 0596521472.
- Ableson, Frank; Collins, Charlie; Sen, Robi (May 1, 2009). *Unlocking Android: A Developer's Guide* (http://www.manning.com/ableson/) (1st ed.). Manning. ISBN 1933988673.
- Conder, Shane; Darcey, Lauren (September 7, 2009). *Android Wireless Application Development* (http://www.informit.com/store/product.aspx?isbn=0321627091) (1st ed.). Addison-Wesley Professional. ISBN 0321627091.
- Murphy, Mark (June 26, 2009). *Beginning Android* (http://www.apress.com/book/view/1430224193) (1st ed.). Apress. ISBN 1430224193.
- Hashimi, Sayed Y.; Komatineni, Satya; MacLean, Dave (February 26, 2010). *Pro Android 2* (http://www.apress.com/book/view/1430226595) (2nd ed.). Apress. ISBN 1430226595.
- Meier, Reto (November 24, 2008). *Professional Android Application Development* (http://www.wrox.com/WileyCDA/WroxTitle/Professional-Android-Application-Development.productCd-0470344717.html) (1st ed.). Wrox Press. ISBN 0470344717.
- DiMarzio, Jerome (July 30, 2008). *Android a programmers guide* (http://www.mhprofessional.com/product.php?isbn=0071599886&cat=112) (1st ed.). McGraw-Hill Osborne Media. ISBN 0071599886.

• Haseman, Chris (July 21, 2008). *Android Essentials* (http://www.apress.com/book/view/1430210648) (1st ed.). Apress. ISBN 1430210648.

External links

- Official Android page (http://www.android.com/)
  - Android Open Source Project (http://source.android.com/)
  - Android Market (http://www.android.com/market)
  - Android Developers (http://developer.android.com/)
  - Android Developers Blog (http://android-developers.blogspot.com/)
  - Android Brand Guidelines (http://www.android.com/branding/)
- A ready-for-Android Eclipse (http://catdroid.org/?p=210&lang=en)
- Google Projects for Android (http://code.google.com/android) from Google Code
- Android Wiki (http://www.androidwiki.com/)
- Sergey Brin introduces the Android platform (http://www.youtube.com/watch?v=1FJHYqE0RDg) at YouTube
- Android: Building a Mobile Platform to Change the Industry (http://www.stanford.edu/class/ee380/Abstracts/071128.html) — lecture given by Google Mobile Platforms Manager, Richard Miner at Stanford University (video archive (http://ee380.stanford.edu/cgi-bin/videologger.php?target=071128-ee380-300.asx)).
- Android (operating system) (http://www.dmoz.org/Computers/Systems/Handhelds/Android/) at the Open Directory Project

GPRS

General packet radio service (GPRS) is a packet oriented mobile data service on the 2G and 3G cellular communication systems global system for mobile communications (GSM). The service is available to users in over 200 countries worldwide. GPRS was originally standardized by European Telecommunications Standards Institute (ETSI) in response to the earlier CDPD and i-mode packet switched cellular technologies. It is now maintained by the 3rd Generation Partnership Project (3GPP).[1] [2] .

It is a best-effort service, as opposed to circuit switching, where a certain quality of service (QoS) is guaranteed during the connection. In 2G systems, GPRS provides data rates of 56-114 kbit/second.[3] 2G cellular technology combined with GPRS is sometimes described as *2.5G*, that is, a technology between the second (2G) and third (3G) generations of mobile telephony[4] . It provides moderate-speed data transfer, by using unused time division multiple access (TDMA) channels in, for example, the GSM system. GPRS is integrated into GSM Release 97 and newer releases.

GPRS usage charging is based on volume of data, either as part of a bundle or on a pay as you use basis. An example of a bundle is up to 5 GB per month for a fixed fee. Usage above the bundle cap is either charged for per megabyte or disallowed. The pay as you use charging is typically per megabyte of traffic. This contrasts with circuit switching data, which is typically billed per minute of connection time, regardless of whether or not the user transfers data during that period.

Technical overview

Services offered

GPRS extends the GSM circuit switched data capabilities and makes the following services possible:

- "Always on" internet access
- Multimedia messaging service (MMS)
- Push to talk over cellular (PoC/PTT)
- Instant messaging and presence—wireless village
- Internet applications for smart devices through wireless application protocol (WAP)
- Point-to-point (P2P) service: inter-networking with the Internet (IP)

If SMS over GPRS is used, an SMS transmission speed of about 30 SMS messages per minute may be achieved. This is much faster than using the ordinary SMS over GSM, whose SMS transmission speed is about 6 to 10 SMS messages per minute.

Protocols supported

GPRS supports the following protocols:

- internet protocol (IP). In practice, mobile built-in browsers use IPv4 since IPv6 is not yet popular.
- point-to-point protocol (PPP). In this mode PPP is often not supported by the mobile phone operator but if the mobile is used as a modem to the connected computer, PPP is used to tunnel IP to the phone. This allows an IP address to be assigned dynamically to the mobile equipment.
- X.25 connections. This is typically used for applications like wireless payment terminals, although it has been removed from the standard. X.25 can still be supported over PPP, or even over IP, but doing this requires either a network based router to perform encapsulation or intelligence built in to the end-device/terminal; e.g., user equipment (UE).

When TCP/IP is used, each phone can have one or more IP addresses allocated. GPRS will store and forward the IP packets to the phone even during handover. The TCP handles any packet loss (e.g. due to a radio noise induced pause).

Hardware

Devices supporting GPRS are divided into three classes:

Class A

Can be connected to GPRS service and GSM service (voice, SMS), using both at the same time. Such devices are known to be available today.

Class B

Can be connected to GPRS service and GSM service (voice, SMS), but using only one or the other at a given time. During GSM service (voice call or SMS), GPRS service is suspended, and then resumed automatically after the GSM service (voice call or SMS) has concluded. Most GPRS mobile devices are Class B.

Class C

Are connected to either GPRS service or GSM service (voice, SMS). Must be switched manually between one or the other service.

A true Class A device may be required to transmit on two different frequencies at the same time, and thus will need two radios. To get around this expensive requirement, a GPRS mobile may implement the dual transfer mode (DTM) feature. A DTM-capable mobile may use simultaneous voice and packet data, with the network coordinating to ensure that it is not required to transmit on two different frequencies at the same time. Such mobiles are considered pseudo-Class A, sometimes referred to as "simple class A". Some networks are expected to support DTM in 2007.

USB 3G/GPRS modems use a terminal-like interface over USB 1.1, 2.0 and later, data formats V.42bis, and RFC 1144 and some models have connector for external antenna. Modems can be added as cards (for laptops) or external USB devices which are similar in shape and size to a computer mouse, or nowadays more like a pendrive.

Huawei E220 3G/GPRS Modem

Addressing

A GPRS connection is established by reference to its access point name (APN). The APN defines the services such as wireless application protocol (WAP) access, short message service (SMS), multimedia messaging service (MMS), and for Internet communication services such as email and World Wide Web access.

In order to set up a GPRS connection for a wireless modem, a user must specify an APN, optionally a user name and password, and very rarely an IP address, all provided by the network operator.

Coding schemes and speeds

The upload and download speeds that can be achieved in GPRS depend on a number of factors such as:

- the number of BTS TDMA time slots assigned by the operator
- the channel encoding used.
- the maximum capability of the mobile device expressed as a GPRS multislot class

Multiple access schemes

The multiple access methods used in GSM with GPRS are based on frequency division duplex (FDD) and TDMA. During a session, a user is assigned to one pair of up-link and down-link frequency channels. This is combined with time domain statistical multiplexing; i.e., packet mode communication, which makes it possible for several users to share the same frequency channel. The packets have constant length, corresponding to a GSM time slot. The down-link uses first-come first-served packet scheduling, while the up-link uses a scheme very similar to reservation ALOHA (R-ALOHA). This means that slotted ALOHA (S-ALOHA) is used for reservation inquiries during a contention phase, and then the actual data is transferred using dynamic TDMA with first-come first-served scheduling.

Channel encoding

Channel encoding is based on a convolutional code at different code rates and GMSK modulation defined for GSM. The following table summarises the options:

Coding scheme	Speed (kbit/s)
CS-1	8.0
CS-2	12.0
CS-3	14.4
CS-4	20.0

The least robust, but fastest, coding scheme (CS-4) is available near a base transceiver station (BTS), while the most robust coding scheme (CS-1) is used when the mobile station (MS) is further away from a BTS.

Using the CS-4 it is possible to achieve a user speed of 20.0 kbit/s per time slot. However, using this scheme the cell coverage is 25% of normal. CS-1 can achieve a user speed of only 8.0 kbit/s per time slot, but has 98% of normal coverage. Newer network equipment can adapt the transfer speed automatically depending on the mobile location.

In addition to GPRS, there are two other GSM technologies which deliver data services: circuit-switched data (CSD) and high-speed circuit-switched data (HSCSD). In contrast to the shared nature of GPRS, these instead establish a dedicated circuit (usually billed per minute). Some applications such as video calling may prefer HSCSD, especially when there is a continuous flow of data between the endpoints.

The following table summarises some possible configurations of GPRS and circuit switched data services.

Technology	Download (kbit/s)	Upload (kbit/s)	TDMA Timeslots allocated
CSD	9.6	9.6	1+1
HSCSD	28.8	14.4	2+1
HSCSD	43.2	14.4	3+1
GPRS	80.0	20.0 (Class 8 & 10 and CS-4)	4+1
GPRS	60.0	40.0 (Class 10 and CS-4)	3+2
EGPRS (EDGE)	236.8	59.2 (Class 8, 10 and MCS-9)	4+1
EGPRS (EDGE)	177.6	118.4 (Class 10 and MCS-9)	3+2

Multislot Class

The multislot class determines the speed of data transfer available in the Uplink and Downlink directions. It is a value between 1 to 45 which the network uses to allocate radio channels in the uplink and downlink direction. Multislot class with values greater than 31 are referred to as high multislot classes.

A multislot allocation is represented as, for example, 5+2. The first number is the number of downlink timeslots and the second is the number of uplink timeslots allocated for use by the mobile station. A commonly used value is class 10 for many GPRS/EGPRS mobiles which uses a maximum of 4 timeslots in downlink direction and 2 timeslots in uplink direction. However simultaneously a maximum number of 5 simultaneous timeslots can be used in both uplink and downlink. The network will automatically configure the for either 3+2 or 4+1 operation depending on the nature of data transfer.

Some high end mobiles, usually also supporting UMTS also support GPRS/EDGE multislot class 32. According to 3GPP TS 45.002 (Release 6), Table B.2, mobile stations of this class support 5 timeslots in downlink and 3 timeslots in uplink with a maximum number of 6 simultaneously used timeslots. If data traffic is concentrated in downlink direction the network will configure the connection for 5+1 operation. When more data is transferred in the uplink the network can at any time change the constellation to 4+2 or 3+3. Under the best reception conditions, i.e. when the best EDGE modulation and coding scheme can be used, 5 timeslots can carry a bandwidth of 5*59.2 kbit/s = 296 kbit/s. In uplink direction, 3 timeslots can carry a bandwidth of 3*59.2 kbit/s = 177.6 kbit/s.[5]

Multislot Classes for GPRS/EGPRS

Multislot Class	Downlink TS	Uplink TS	Active TS
1	1	1	2
2	2	1	3
3	2	2	3
4	3	1	4
5	2	2	4
6	3	2	4
7	3	3	4
8	4	1	5
9	3	2	5
10	4	2	5
11	4	3	5
12	4	4	5
30	5	1	6
31	5	2	6
32	5	3	6
33	5	4	6
34	5	5	6

Attributes of a multislot class

Each multislot class identifies the following:

- the maximum number of Timeslots that can be allocated on uplink
- the maximum number of Timeslots that can be allocated on downlink
- the total number of timeslots which can be allocated by the network to the mobile
- the time needed for the mobile phone to perform adjacent cell signal level measurement and get ready to transmit
- the time needed for the MS to get ready to transmit
- the time needed for the MS to perform adjacent cell signal level measurement and get ready to receive
- the time needed for the MS to get ready to receive.

The different multislot class specification is detailed in the Annex B of the 3GPP Technical Specification 45.002 (Multiplexing and multiple access on the radio path)

Usability

The maximum speed of a GPRS connection offered in 2003 was similar to a modem connection in an analog wire telephone network, about 32-40 kbit/s, depending on the phone used. Latency is very high; round-trip time (RTT) is typically about 600-700 ms and often reaches 1 s. GPRS is typically prioritized lower than speech, and thus the quality of connection varies greatly.

Devices with latency/RTT improvements (via, for example, the extended UL TBF mode feature) are generally available. Also, network upgrades of features are available with certain operators. With these enhancements the active round-trip time can be reduced, resulting in significant increase in application-level throughput speeds.

See also

- Code division multiple access (CDMA)
- Enhanced data rates for GSM evolution (EDGE)
- Universal mobile telephone system (UMTS)
- GPRS core network
- Sub-network dependent convergence protocol (SNDCP)
- IP Multimedia Subsystem
- High-speed downlink packet access (HSDPA)
- List of device bandwidths

References

[1] http://www.etsi.org/WebSite/homepage.aspx ETSI
[2] http://www.3gpp.org/3GPP
[3] http://about.qkport.com/g/general_packet_radio_service General packet radio service from Qkport
[4] http://www.funsms.net/mobile_phone_generations.htm Mobile Phone Generations from
[5] http://mobilesociety.typepad.com/mobile_life/2007/04/gprs_and_edge_m.html

External links

- 3GPP AT command set for user equipment (UE) (http://www.3gpp.org/ftp/Specs/latest/Rel-8/27_series/27007-841.zip)
- GPRS security information (archive.org) (http://web.archive.org/web/20080209213430/http://www.gprssecurity.com/)
- Free GPRS resources (http://www.telecomspace.com/datatech-gprs.html)
- Free online tutorial (http://www.comsoc.org/livepubs/surveys/public/3q99issue/bettstetter.html).
- GSM World, the trade association for GSM and GPRS network operators (http://www.gsmworld.com/technology/gprs/intro.shtml).
- Palowireless GPRS resource center (http://www.palowireless.com/gprs/)
- GPRS attach and PDP context activation sequence diagram (http://www.eventhelix.com/RealtimeMantra/Telecom/gprs_attach_pdp_sequence_diagram.pdf)

UMTS

Universal Mobile Telecommunications System (UMTS) is one of the third-generation (3G) mobile telecommunications technologies, which is also being developed into a 4G technology. The first deployment of the UMTS is the **release99** (R99) architecture. It is specified by 3GPP and is part of the global ITU IMT-2000 standard. The most common form of UMTS uses W-CDMA (IMT Direct Spread) as the underlying air interface but the system also covers TD-CDMA and TD-SCDMA (both IMT CDMA TDD). Being a complete network system, UMTS also covers the radio access network (UMTS Terrestrial Radio Access Network, or UTRAN) and the core network (Mobile Application Part, or MAP), as well as authentication of users via USIM cards (Subscriber Identity Module).

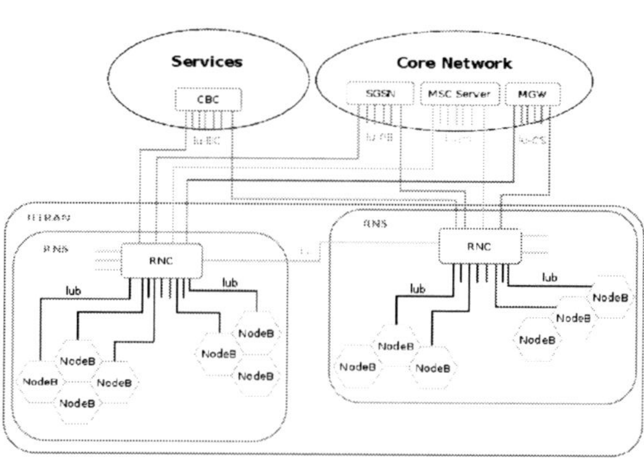

UMTS Network Architecture

Unlike EDGE (IMT Single-Carrier, based on GSM) and CDMA2000 (IMT Multi-Carrier), UMTS requires new base stations and new frequency allocations. However, it is closely related to GSM/EDGE as it borrows and builds upon concepts from GSM. Further, most UMTS handsets also support GSM, allowing seamless dual-mode operation. Therefore, UMTS is sometimes marketed as **3GSM**, emphasizing the close relationship with GSM and differentiating it from competing technologies.

The name UMTS, introduced by ETSI, is usually used in Europe. Outside of Europe, the system is also known by other names such as **FOMA**[1] or **W-CDMA**.[2] [1] In marketing, it is often referred to as **3G** or **3G+**.

Features

UMTS, using 3GPP, supports maximum theoretical data transfer rates of 42 Mbit/s (with HSPA+),[3] although at the moment users in deployed networks can expect a transfer rate of up to 384 kbit/s for R99 handsets, and 7.2 Mbit/s for HSDPA handsets in the downlink connection. This is still much greater than the 9.6 kbit/s of a single GSM error-corrected circuit switched data channel or multiple 9.6 kbit/s channels in HSCSD (14.4 kbit/s for CDMAOne), and—in competition to other network technologies such as CDMA2000, PHS or WLAN—offers access to the World Wide Web and other data services on mobile devices.

Precursors to 3G are 2G mobile telephony systems, such as GSM, IS-95, PDC, CDMA PHS and other 2G technologies deployed in different countries. In the case of GSM, there is an evolution path from 2G, to GPRS, also known as 2.5G. GPRS supports a much better data rate (up to a theoretical maximum of 140.8 kbit/s, though typical rates are closer to 56 kbit/s) and is packet switched rather than connection oriented (circuit switched). It is deployed in many places where GSM is used. E-GPRS, or EDGE, is a further evolution of GPRS and is based on more modern coding schemes. With EDGE the actual packet data rates can reach around 180 kbit/s (effective). EDGE systems are often referred as "2.75G Systems".

Since 2006, UMTS networks in many countries have been or are in the process of being upgraded with High Speed Downlink Packet Access (HSDPA), sometimes known as 3.5G. Currently, HSDPA enables downlink transfer speeds of up to 21 Mbit/s. Work is also progressing on improving the uplink transfer speed with the High-Speed Uplink Packet Access (HSUPA). Longer term, the 3GPP Long Term Evolution project plans to move UMTS to 4G speeds of 100 Mbit/s down and 50 Mbit/s up, using a next generation air interface technology based upon Orthogonal frequency-division multiplexing.

The first national consumer UMTS networks launched in 2002 with a heavy emphasis on telco-provided mobile applications such as mobile TV and video calling. The high data speeds of UMTS are now most often utilised for Internet access: experience in Japan and elsewhere has shown that user demand for video calls is not high, and telco-provided audio/video content has declined in popularity in favour of high-speed access to the World Wide Web - either directly on a handset or connected to a computer via Wi-Fi, Bluetooth, Infrared or USB.

Technology

UMTS combines three different air interfaces, GSM's Mobile Application Part (MAP) core, and the GSM family of speech codecs.

Air interfaces

UMTS provides several different terrestrial air interfaces, called **UMTS Terrestrial Radio Access** (**UTRA**).[4] All air interface options are part of ITU's IMT-2000. In the currently most popular variant for cellular mobile telephones, W-CDMA (IMT Direct Spread) is used.

Please note that the terms W-CDMA, TD-CDMA and TD-SCDMA are misleading. While they suggest covering just a channel access method (namely a variant of CDMA), they are actually the common names for the whole air interface standards.[5]

Non-terrestrial radio access networks are currently under research.

W-CDMA (UTRA-FDD)

W-CDMA uses the DS-CDMA channel access method with a pair of 5 MHz channels. In contrast, the competing CDMA2000 system uses one or more arbitrary 1.25 MHz channels for each direction of communication. W-CDMA systems are widely criticized for their large spectrum usage, which has delayed deployment in countries that acted relatively slowly in allocating new frequencies specifically for 3G services (such as the United States).

UMTS transmitter on the roof of a building

The specific frequency bands originally defined by the UMTS standard are 1885–2025 MHz for the mobile-to-base (uplink) and 2110–2200 MHz for the base-to-mobile (downlink). In the US, 1710–1755 MHz and 2110–2155 MHz will be used instead, as the 1900 MHz band was already used.[6] While UMTS2100 is the most widely-deployed UMTS band, some countries' UMTS operators use the 850 MHz and/or 1900 MHz bands (independently, meaning uplink and downlink are within the same band), notably in the US by AT&T Mobility, New Zealand by Telecom New Zealand on the XT Mobile Network and in Australia by Telstra on the Next G network.

W-CDMA is a part of IMT-2000 as **IMT Direct Spread**.

UTRA-TDD HCR

UMTS-TDD's air interfaces that use the TD-CDMA channel access technique are standardized as **UTRA-TDD HCR**, which uses increments of 5 MHz of spectrum, each slice divided into 10ms frames containing fifteen time slots (1500 per second)[7] . The time slots (TS) are allocated in fixed percentage for downlink and uplink. TD-CDMA is used to multiplex streams from or to multiple transceivers. Unlike W-CDMA, it does not need separate frequency bands for up- and downstream, allowing deployment in tight frequency bands.

TD-CDMA is a part of IMT-2000 as **IMT CDMA TDD**.

TD-SCDMA (UTRA-TDD 1.28 Mcps Low Chip Rate)

TD-SCDMA uses the TDMA channel access method combined with an adaptive synchronous CDMA component [8] on 1.6 MHz slices of spectrum, allowing deployment in even tighter frequency bands than TD-CDMA. However, the main incentive for development of this Chinese-developed standard was avoiding or reducing the license fees that have to be paid to non-Chinese patent owners. Unlike the other air interfaces, TD-SCDMA was not part of UMTS from the beginning but has been added in Release 4 of the specification.

Like TD-CDMA, it is known as **IMT CDMA TDD** within IMT-2000.

Radio access network

UMTS also specifies the **UMTS Terrestrial Radio Access Network** (**UTRAN**), which is composed of multiple base stations, possibly using different terrestrial air interface standards and frequency bands.

UMTS and GSM/EDGE can share a Core Network (CN), making UTRAN an alternative radio access network to GERAN (GSM/EDGE RAN), and allowing (mostly) transparent switching between the RANs according to available coverage and service needs. Because of that, UMTS' and GSM/EDGE's radio access networks are sometimes collectively referred to as **UTRAN/GERAN**.

UMTS networks are often combined with GSM/EDGE, the later of which is also a part of IMT-2000.

The UE (User Equipment) interface of the RAN (Radio Access Network) primarily consists of RRC (Radio Resource Control), RLC (Radio Link Control) and MAC (Media Access Control) protocols. RRC protocol handles connection establishment, measurements, radio bearer services, security and handover decisions. RLC protocol primarily divides into three Modes - Transparent Mode (TM), Unacknowledge Mode (UM), Acknowledge Mode (AM). The functionality of AM entity resembles TCP operation where as UM operation resembles UDP operation. In TM mode, data will be sent to lower layers without adding any header to SDU of higher layers. MAC handles the scheduling of data on air interface depending on higher layer (RRC) configured parameters.

Set of properties related to data transmission is called Radio Bearer (RB). This set of properties will decide the maximum allowed data in a TTI (Transmission Time Interval). RB includes RLC information and RB mapping. RB mapping decides the mapping between RB<->logical channel<->transport channel. Signaling message will be send on Signaling Radio Bearers (SRBs) and data packets (either CS or PS) will be sent on data RBs. RRC and NAS messages will go on SRBs.

Security includes two procedures: integrity and ciphering. Integrity validates the resource of message and also make sure that no one (third/unknown party) on radio interface has not modified message. Ciphering make sure that no one listens your data on air interface. Both integrity and ciphering will be applied for SRBs where as only ciphering will be applied for data RBs.

Core network

With **Mobile Application Part**, UMTS uses the same core network standard as GSM/EDGE. This allows a simple migration for existing GSM operators. However, the migration path to UMTS is still costly: while much of the core infrastructure is shared with GSM, the cost of obtaining new spectrum licenses and overlaying UMTS at existing towers is high.

The CN can be connected to various backbone networks like the Internet, ISDN. UMTS (and GERAN) include the three lowest layers of OSI model. The network layer (OSI 3) includes the Radio Resource Management protocol (RRM) that manages the bearer channels between the mobile terminals and the fixed network, including the handovers. abc

Spectrum allocation

Over 130 licenses have already been awarded to operators worldwide (as of December 2004), specifying W-CDMA radio access technology that builds on GSM. In Europe, the license process occurred at the tail end of the technology bubble, and the auction mechanisms for allocation set up in some countries resulted in some extremely high prices being paid for the original 2100 MHz licenses, notably in the UK and Germany. In Germany, bidders paid a total €50.8 billion for six licenses, two of which were subsequently abandoned and written off by their purchasers (Mobilcom and the Sonera/Telefonica consortium). It has been suggested that these huge license fees have the character of a very large tax paid on future income expected many years down the road. In any event, the high prices paid put some European telecom operators close to bankruptcy (most notably KPN). Over the last few years some operators have written off some or all of the license costs. Between 2007..2009 all three Finnish carriers begun to use 900 MHz UMTS in a shared arrangement with its surrounding 2G GSM base stations for rural area coverage, a trend that is expected to expand over Europe in the next 1–3 years.

The 2100 MHz UMTS spectrum allocated in Europe is already used in North America. The 1900 MHz range is used for 2G (PCS) services, and 2100 MHz range is used for satellite communications. Regulators have, however, freed up some of the 2100 MHz range for 3G services, together with the 1700 MHz for the uplink. UMTS operators in North America who want to implement a European style 2100/1900 MHz system will have to share spectrum with existing 2G services in the 1900 MHz band.

AT&T Wireless launched UMTS services in the United States by the end of 2004 strictly using the existing 1900 MHz spectrum allocated for 2G PCS services. Cingular acquired AT&T Wireless in 2004 and has since then launched UMTS in select US cities. Cingular renamed itself AT&T and is rolling out some cities with a UMTS network at 850 MHz to enhance its existing UMTS network at 1900 MHz and now offers subscribers a number of UMTS 850/1900 phones.

T-Mobile's rollout of UMTS in the US will focus on the 2100/1700 MHz bands.

In Canada, UMTS coverage is being provided on the 850 MHz and 1900 MHz band on the Rogers, Bell, and Telus networks. Recently, new providers Wind Mobile and Mobilicity, have begun operations in the 2100/1700 MHz bands and Quebecor and Shaw Communications are planning their own launches in coming years.

In 2008, Australian telco Telstra replaced its existing CDMA network with a national 3G network, branded as NextG, operating in the 850 MHz band. Telstra currently provides UMTS service on this network, and also on the 2100 MHz UMTS network, through a co-ownership of the owning and administrating company 3GIS. This company is also co-owned by Hutchison 3G Australia, and this is the primary network used by their customers. Optus is currently rolling out a 3G network operating on the 2100 MHz band in cities and most large towns, and the 900 MHz band in regional areas. Vodafone is also building a 3G network using the 900 MHz band.

In India BSNL has started its 3G services since October 2009 beginning with the larger cities and then expanding over to smaller cities. The 850 MHz and 900 MHz bands provide greater coverage compared to equivalent 1700/1900/2100 MHz networks, and are best suited to regional areas where greater distances separate subscriber and base station.

Carriers in South America are now also rolling out 850 MHz networks.

Interoperability and global roaming

UMTS phones (and data cards) are highly portable—they have been designed to roam easily onto other UMTS networks (if the providers have roaming agreements in place). In addition, almost all UMTS phones are UMTS/GSM dual-mode devices, so if a UMTS phone travels outside of UMTS coverage during a call the call may be transparently handed off to available GSM coverage. Roaming charges are usually significantly higher than regular usage charges.

Most UMTS licensees consider ubiquitous, transparent global roaming an important issue. To enable a high degree of interoperability, UMTS phones usually support several different frequencies in addition to their GSM fallback. Different countries support different UMTS frequency bands – Europe initially used 2100 MHz while the most carriers in the USA use 850Mhz and 1900Mhz. T-mobile has launched a network in the US operating at 1700 MHz (uplink) /2100 MHz (downlink), and these bands are also being adopted elsewhere in the Americas. A UMTS phone and network must support a common frequency to work together. Because of the frequencies used, early models of UMTS phones designated for the United States will likely not be operable elsewhere and vice versa. There are now 11 different frequency combinations used around the world—including frequencies formerly used solely for 2G services.

UMTS phones can use a Universal Subscriber Identity Module, USIM (based on GSM's SIM) and also work (including UMTS services) with GSM SIM cards. This is a global standard of identification, and enables a network to identify and authenticate the (U)SIM in the phone. Roaming agreements between networks allow for calls to a customer to be redirected to them while roaming and determine the services (and prices) available to the user. In addition to user subscriber information and authentication information, the (U)SIM provides storage space for phone book contact. Handsets can store their data on their own memory or on the (U)SIM card (which is usually more limited in its phone book contact information). A (U)SIM can be moved to another UMTS or GSM phone, and the phone will take on the user details of the (U)SIM, meaning it is the (U)SIM (not the phone) which determines the phone number of the phone and the billing for calls made from the phone.

Japan was the first country to adopt 3G technologies, and since they had not used GSM previously they had no need to build GSM compatibility into their handsets and their 3G handsets were smaller than those available elsewhere. In 2002, NTT DoCoMo's FOMA 3G network was the first commercial UMTS network—using a pre-release specification[9] , it was initially incompatible with the UMTS standard at the radio level but used standard USIM cards, meaning USIM card based roaming was possible (transferring the USIM card into a UMTS or GSM phone when travelling). Both NTT DoCoMo and SoftBank Mobile (which launched 3G in December 2002) now use standard UMTS.

Handsets and modems

All of the major 2G phone manufacturers (that are still in business) are now manufacturers of 3G phones. The early 3G handsets and modems were specific to the frequencies required in their country, which meant they could only roam to other countries on the same 3G frequency (though they can fall back to the older GSM standard). Canada and USA have a common share of frequencies, as do most European countries. The article UMTS frequency bands is an overview of UMTS network frequencies around the world.

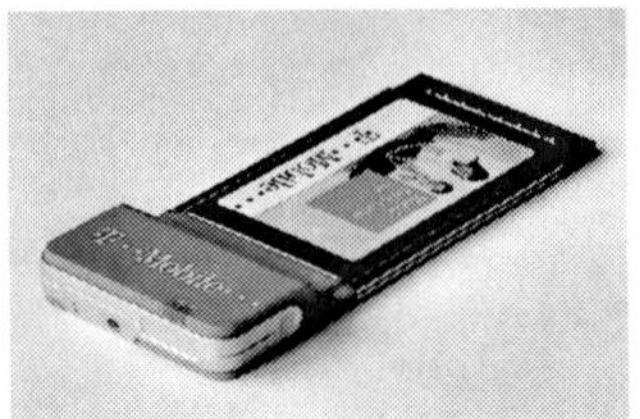

T-Mobile UMTS PC Card modem

Using a cellular router, PCMCIA or USB card, customers are able to access 3G broadband services, regardless of their choice of computer (such as a tablet PC or a PDA). Some software installs itself from the modem, so that in some cases absolutely no knowledge of technology is required to get online in moments. Using a phone that supports 3G and Bluetooth 2.0, multiple Bluetooth-capable laptops can be connected to the Internet. Some smartphones can also act as a mobile WLAN access point.

The Nokia 6650, an early UMTS handset

There are almost no 3G phones or modems available supporting all 3G frequencies (UMTS850/900/1700/1900/2100 MHz). However, many phones are offering more than one band which still enables extensive roaming. For example, a tri-band chipset operating on 850/1900/2100 MHz, such as that found in Apple's iPhone, allows usage in the majority of countries where UMTS-FDD is deployed.

Other competing standards

The main competitor to UMTS is CDMA2000 (IMT-MC), which is developed by the 3GPP2. Unlike UMTS, CDMA2000 is an evolutionary upgrade to an existing 2G standard, cdmaOne, and is able to operate within the same frequency allocations. This and CDMA2000's narrower bandwidth requirements make it easier to deploy in existing spectra. In some, but not all, cases, existing GSM operators only have enough spectrum to implement either UMTS or GSM, not both. For example, in the US D, E, and F PCS spectrum blocks [10], the amount of spectrum available is 5 MHz in each direction. A standard UMTS system would saturate that spectrum. Where CDMA2000 is deployed, it usually co-exists with UMTS. In many markets however, the co-existence issue is of little relevance, as legislative hurdles exist to co-deploying two standards in the same licensed slice of spectrum.

Another competitor to UMTS is EDGE (IMT-SC), which is an evolutionary upgrade to the 2G GSM system, leveraging existing GSM spectrums. It is also much easier, quicker, and considerably cheaper for wireless carriers to "bolt-on" EDGE functionality by upgrading their existing GSM transmission hardware to support EDGE than having to install almost all brand-new equipment to deliver UMTS. However, being developed by 3GPP just as UMTS, EDGE is not a true competitor. Instead, it is used as a temporary solution preceding UMTS roll-out or as a complement for rural areas. This is facilitated by the fact that GSM/EDGE and UMTS specification are jointly developed and rely on the same core network, allowing dual-mode operation including vertical handovers.

China's TD-SCDMA standard is often seen as a competitor, too. TD-SCDMA has been added to UMTS' Release 4 as UTRA-TDD 1.28 Mcps Low Chip Rate (UTRA-TDD LCR). Unlike TD-CDMA (UTRA-TDD 3.84 Mcps High Chip Rate, UTRA-TDD HCR) which complements W-CDMA (UTRA-FDD), it is suitable for both micro and macro cells. However, the lack of vendors' support is preventing it from being a real competitor.

While DECT is technically capable of competing with UMTS and other cellular networks in densely-populated, urban areas, it has only been deployed for domestic cordless phones and private in-house networks.

All of these competitors have been accepted by ITU as part of the IMT-2000 family of 3G standards, along with UMTS-FDD.

On the Internet access side, competing systems include WiMAX and Flash-OFDM.

Migrating from GPRS to UMTS

From GPRS network, the following network elements can be reused:

- Home Location Register (HLR)
- Visitor Location Register (VLR)
- Equipment Identity Register (EIR)
- Mobile Switching Center (MSC) (vendor dependent)
- Authentication Center (AUC)
- Serving GPRS Support Node (SGSN) (vendor dependent)
- Gateway GPRS Support Node (GGSN)

From Global Service for Mobile (GSM) communication radio network, the following elements cannot be reused

- **Base station controller (BSC)**
- Base transceiver station (BTS)

They can remain in the network and be used in dual network operation where 2G and 3G networks co-exist while network migration and new 3G terminals become available for use in the network.

The UMTS network introduces new network elements that function as specified by 3GPP:

- Node B (base transceiver station)
- Radio Network Controller (RNC)
- Media Gateway (MGW)

The functionality of MSC and SGSN changes when going to UMTS. In a GSM system the MSC handles all the circuit switched operations like connecting A- and B-subscriber through the network. SGSN handles all the packet switched operations and transfers all the data in the network. In UMTS the Media gateway (MGW) take care of all data transfer in both circuit and packet switched networks. MSC and SGSN control MGW operations. The nodes are renamed to MSC-server and GSN-server.

Problems and issues

Some countries, including the United States and Japan, have allocated spectrum differently from the ITU recommendations, so that the standard bands most commonly used for UMTS (UMTS-2100) have not been available. In those countries, alternative bands are used, preventing the interoperability of existing UMTS-2100 equipment, and requiring the design and manufacture of different equipment for the use in these markets. As is the case with GSM900 today, standard UMTS 2100 MHz equipment will not work in those markets, However, it appears as though UMTS is not suffering as much from handset band compatibility issues as GSM did, as many UMTS handsets are multi-band in both UMTS and GSM modes. Quad-band GSM (850, 900, 1800, and 1900 MHz bands) and tri-band UMTS (850, 1900, and 2100 MHz bands) handsets are becoming more commonplace.

The early days of UMTS saw rollout hitches in many countries. Overweight handsets with poor battery life were first to arrive on a market highly sensitive to weight and form factor. The Motorola A830, a debut handset on Hutchison's 3 network, weighed more than 200 grams and even featured a detachable camera to reduce handset weight. Another significant issue involved call reliability, related to problems with handover from UMTS to GSM. Customers found their connections being dropped as handovers were possible only in one direction (UMTS → GSM), with the handset only changing back to UMTS after hanging up. In most networks around the world this is no longer an issue.

Compared to GSM, UMTS networks initially required a higher base station density. For fully-fledged UMTS incorporating video on demand features, one base station needed to be set up every 1–1.5 km (0.62–0.93 mi). This was the case when only the 2100 MHz band was being used, however with the growing use of lower-frequency bands (such as 850 and 900 MHz) this is no longer so. This has led to increasing rollout of the lower-band networks by operators since 2006.

Even with current technologies and low-band UMTS, telephony and data over UMTS is still more power intensive than on comparable GSM networks. Apple, Inc. cited[11] UMTS power consumption as the reason that the first generation iPhone only supported EDGE. Their release of the iPhone 3G quotes talk time on UMTS as half that available when the handset is set to use GSM. Other manufacturers indicate different battery life time for UMTS mode compared to GSM mode as well. As battery and network technology improves, this issue is diminishing.

Releases

The evolution of UMTS progresses according to planned releases. Each release is designed to introduce new features and improve upon existing ones.

Release '99

- Bearer services
- 64 kbit/s circuit switch
- 384 kbit/s packet switched
- Location services
- Call services: compatible with Global System for Mobile Communications (GSM), based on Universal Subscriber Identity Module (USIM)

Release 4

- Edge radio
- Multimedia messaging
- MExE (Mobile Execution Environment)
- Improved location services
- IP Multimedia Services (IMS)

Release 5

- IP Multimedia Subsystem (IMS)
- IPv6, IP transport in UTRAN
- Improvements in GERAN, MExE, etc
- HSDPA

Release 6

- WLAN integration
- Multimedia broadcast and multicast
- Improvements in IMS
- HSUPA
- Fractional DPCH

Release 7

- Enhanced L2
- 64 QAM , MIMO
- VoIP over HSPA
- CPC - continuous packet connectivity
- FRLC - Flexible RLC

Release 8

- DC-HSPA
- HSUPA 16QAM

See also

- List of Deployed UMTS networks
- 3G
- 3GPP: the body that manages the UMTS standard.
- 3GPP Long Term Evolution, the 3GPP project to evolve UMTS towards 4G capabilities.
- LSTI
- GAN/UMA: A standard for running GSM and UMTS over wireless LANs.
- Opportunity Driven Multiple Access, ODMA: a UMTS TDD mode communications relaying protocol
- HSDPA, HSUPA: updates to the W-CDMA air interface.
- PDCP
- Subscriber Identity Module
- UMTS-TDD: a variant of UMTS largely used to provide wireless Internet service.
- UMTS frequency bands
- W-CDMA: the primary air interface standard used by UMTS.
- W-CDMA 2100

Other, non-UMTS, 3G and 4G standards:

- CDMA2000: evolved from the cmdaOne (also known as IS-95, or "CDMA") standard, managed by the 3GPP2
- FOMA
- TD-SCDMA
- WiMAX: a newly emerging wide area wireless technology.

UMTS is an evolution of the GSM mobile phone standard.

- GSM
- GPRS
- EDGE
- ETSI

Other useful information

- Mobile modem
- Spectral efficiency comparison table
- Code Division Multiple Access (CDMA)
- Common pilot channel or CPICH, a simple synchronisation channel in WCDMA.
- Multiple-input multiple-output (MIMO) is the major issue of multiple antenna research.
- Wi-Fi: a local area wireless technology that is complementary to UMTS.
- List of device bandwidths
- Operations and Maintenance Centre

- Radio Network Controller
- UMTS security

Literature

- Martin Sauter: *Communication Systems for the Mobile Information Society*, John Wiley, September 2006, ISBN 0-470-02676-6
- Ahonen and Barrett (editors), *Services for UMTS* (Wiley, 2002) first book on the services for 3G, ISBN 978-0-471-48550-6
- Holma and Toskala (editors), *WCDMA for UMTS*, (Wiley, 2000) first book dedicated to 3G technology, ISBN 978-0-471-72051-5
- Kreher and Ruedebusch, *UMTS Signaling: UMTS Interfaces, Protocols, Message Flows and Procedures Analyzed and Explained* (Wiley 2007), ISBN 978-0-470-06533-4
- Laiho, Wacker and Novosad, *Radio Network Planning and Optimization for UMTS* (Wiley, 2002) first book on radio network planning for 3G, ISBN 978-0-470-01575-9

Notes

[1] 3GPP notes that "there currently existed many different names for the same system (eg FOMA, W-CDMA, UMTS, etc)"; 3GPP. "Draft summary minutes, decisions and actions from 3GPP Organizational Partners Meeting#6, Tokyo, 9 October 2001" (http://www.3gpp.org/ftp/op/OP_07/DOCS/pdf/OP6_13r1.pdf) (PDF). pp. 7. .

[2] The term W-CDMA usually refers to UMTS' main air interface, UTRA-FDD, or networks which only operate on UTRA-FDD. However, there are rare instances where it is used in a broader sense, as a synonym for UMTS or any UMTS air interface. For example, 3GPP refers to "[b]oth Frequency Division Duplex (FDD) and Time Division Duplex (TDD) variants" of W-CDMA,3GPP. "Keywords (WCDMA, HSPA, LTE, etc): W-CDMA" (http://www.3gpp.org/article/w-cdma). . Retrieved 2009-06-15. i.e. UTRA-FDD and UTRA-TDD.

[3] Tindal, Suzanne (8 December 2008). "Telstra boosts Next G to 21Mbps" (http://www.zdnet.com.au/news/communications/soa/Telstra-boosts-Next-G-to-21Mbps/0,130061791,339293706,00.htm). ZDNet Australia. . Retrieved 2009-03-16.

[4] 3GNewsroom.com (2003-11-29). "3G Glossary - UTRA" (http://www.3gnewsroom.com/html/glossary/u.shtml). . Retrieved 2009-02-16.

[5] ITU-D Study Group 2. "Guidelines on the smooth transition of existing mobile networks to IMT-2000 for developing countries (GST); Report on Question 18/2" (http://www.itu.int/dms_pub/itu-d/opb/stg/D-STG-SG02.18-1-2006-PDF-E.pdf). pp. 4, 25–28. . Retrieved 2009-06-15.

[6] The FCC's Advanced Wireless Services bandplan (http://wireless.fcc.gov/services/aws/data/awsbandplan.pdf)

[7] Forkel et al. (2002). "Performance Comparison Between UTRA-TDD High Chip Rate And Low Chip Rate Operation" (http://citeseerx.ist.psu.edu/viewdoc/summary?doi=10.1.1.11.3672). . Retrieved 2009-02-16.

[8] Siemens (2004-06-10). "TD-SCDMA Whitepaper: the Solution for TDD bands" (http://www.tdscdma-forum.org/en/pdfword/200511817463050335.pdf) (pdf). TD Forum. pp. 6–9. . Retrieved 2009-06-15.

[9] Hsiao-Hwa Chen (2007), John Wiley and Sons, pp. 105–106, ISBN 978-047002294-8

[10] http://wireless.fcc.gov/auctions/data/bandplans/pcsband.pdf

[11] iPhone 'Surfing' On AT&T Network Isn't Fast, Jobs Concedes (http://online.wsj.com/article/SB118306134626851922.html)

References

External links

- 3GPP Specifications Numbering Schemes (http://www.3gpp.org/specs/numbering.htm)
- Vocabulary for 3GPP Specifications, up to Release 8 (http://www.3gpp.org/ftp/Specs/html-info/21905.htm)
- UMTS FAQ (http://www.umtsworld.com/umts/faq.htm) on UMTS World
- Worldwide W-CDMA frequency allocations (http://www.umtsworld.com/technology/frequencies.htm) on UMTS World
- UMTS TDD Alliance (http://www.umtstdd.org) The Global UMTS TDD Alliance
- 3GSM World Congress (http://www.3gsmworldcongress.com/)
- UMTS Provider Chart (http://www.spectran.com/Frequenzplan-UMTS_en.shtml)

HSDPA

High-Speed Downlink Packet Access (**HSDPA**) is an enhanced 3G (third generation) mobile telephony communications protocol in the High-Speed Packet Access (HSPA) family, also dubbed 3.5G, 3G+ or turbo 3G, which allows networks based on Universal Mobile Telecommunications System (UMTS) to have higher data transfer speeds and capacity. Current HSDPA deployments support down-link speeds of 1.8, 3.6, 7.2 and 14.0 Mbit/s. Further speed increases are available with HSPA+, which provides speeds of up to 42 Mbit/s downlink and 84 Mbit/s with Release 9 of the 3GPP standards.[1]

Technology

HS-DSCH channel

For HSDPA, a new transport layer channel, High-Speed Downlink Shared Channel (HS-DSCH), has been added to W-CDMA release 5 and further specification. It is implemented by introducing three new physical layer channels: HS-SCCH, HS-DPCCH and HS-PDSCH. The High Speed-Shared Control Channel (HS-SCCH) informs the user that data will be sent on the HS-DSCH 2 slots ahead. The Uplink High Speed-Dedicated Physical Control Channel (HS-DPCCH) carries acknowledgment information and current channel quality indicator (CQI) of the user. This value is then used by the base station to calculate how much data to send to the user devices on the next transmission. The High Speed-Physical Downlink Shared Channel (HS-PDSCH) is the channel mapped to the above HS-DSCH transport channel that carries actual user data.

Hybrid automatic repeat-request (HARQ)

Data is transmitted together with error correction bits. Minor errors can thus be corrected without retransmission; see forward error correction.

If retransmission is needed, the user device saves the packet and later combines it with retransmitted packet to recover the error-free packet as efficiently as possible. Even if the retransmitted packets are corrupted, their combination can yield an error-free packet. Retransmitted packet may be either identical (chase combining) or different from the first transmission (incremental redundancy).

The round-trip time for retransmissions is improved since the retransmissions are done from base station instead of radio network controller.

Fast packet scheduling

The HS-DSCH downlink channel is shared between users using channel-dependent scheduling to make the best use of available radio conditions. Each user device continually transmits an indication of the downlink signal quality, as often as 500 times per second. Using this information from all devices, the base station decides which users will be sent data on the next 2 ms frame and how much data should be sent for each user. More data can be sent to users which report high downlink signal quality.

The amount of the channelisation code tree, and thus network bandwidth, allocated to HSDPA users is determined by the network. The allocation is "semi-static" in that it can be modified while the network is operating, but not on a frame-by-frame basis. This allocation represents a trade-off between bandwidth allocated for HSDPA users, versus that for voice and non-HSDPA data users. The allocation is in units of channelisation codes for Spreading Factor 16, of which 16 exist and up to 15 can be allocated to HSDPA. When the base station decides which users will receive data on the next frame, it also decides which channelisation codes will be used for each user. This information is sent to the user devices over one or more "scheduling channels"; these channels are not part of the HSDPA allocation previously mentioned, but are allocated separately. Thus, for a given 2 ms frame, data may be sent to a number of users simultaneously, using different channelisation codes. The maximum number of users to receive data on a given 2 ms frame is determined by the number of allocated channelisation codes. By contrast, in CDMA2000 1xEV-DO, data is sent to only one user at a time.

Adaptive modulation and coding

The modulation scheme and coding are changed on a per-user basis, depending on signal quality and cell usage. The initial scheme is Quadrature phase-shift keying (QPSK), but in good radio conditions 16QAM and 64QAM can significantly increase data throughput rates. With 5 Code allocation, QPSK typically offers up to 1.8 Mbit/s peak data rates, while 16QAM offers up to 3.6. Additional codes (e.g. 10, 15) can also be used to improve these data rates or extend the network capacity throughput significantly.

Other improvements

HSDPA is part of the UMTS standards since release 5, which also accompanies an improvement on the uplink providing a new bearer of 384 kbit/s. The previous maximum bearer was 128 kbit/s.

As well as improving data rates, HSDPA also decreases latency and so the round trip time for applications.

In later 3GPP specification releases HSPA+ increases data rates further by adding 64QAM modulation, MIMO and Dual-Cell HSDPA operation, i.e. two 5 MHz carriers are used simultaneously.

HSDPA User Equipment (UE) categories

HSDPA comprises various versions with different data speeds. The following table is derived from table 5.1a of the release 9 version of 3GPP TS 25.306 [2] and shows maximum speeds of different device classes and by what combination of features they are achieved. In 2009 the most common devices are category 6 (3.6 Mbit/s) and category 8 (7.2 Mbit/s) with retail prices around 60 euros without subscription.

Protocol	3GPP Release	Category	Max. number of HS-DSCH codes	Modulation	MIMO, Dual-Cell	Code rate at max. data rate[3]	Max. data rate [Mbit/s]
HSDPA	Release 5	1	5	16-QAM		.76	1.2
HSDPA	Release 5	2	5	16-QAM		.76	1.2
HSDPA	Release 5	3	5	16-QAM		.76	1.8
HSDPA	Release 5	4	5	16-QAM		.76	1.8
HSDPA	Release 5	5	5	16-QAM		.76	3.6
HSDPA	Release 5	6	5	16-QAM		.76	3.6
HSDPA	Release 5	7	10	16-QAM		.75	7.2
HSDPA	Release 5	8	10	16-QAM		.76	7.2
HSDPA	Release 5	9	15	16-QAM		.70	10.1
HSDPA	Release 5	10	15	16-QAM		.97	14.0
HSDPA	Release 5	11	5	QPSK		.76	0.9
HSDPA	Release 5	12	5	QPSK		.76	1.8
HSPA+	Release 7	13	15	64-QAM		.82	17.6
HSPA+	Release 7	14	15	64-QAM		.98	21.1
HSPA+	Release 7	15	15	16-QAM	MIMO	.81	23.4
HSPA+	Release 7	16	15	16-QAM	MIMO	.97	28.0

HSPA+	Release 7	19	15	64-QAM	MIMO	.82	35.3
HSPA+	Release 7	20	15	64-QAM	MIMO	.98	42.2
Dual-Cell HSDPA	Release 8	21	15	16-QAM	Dual-Cell	.81	23.4
Dual-Cell HSDPA	Release 8	22	15	16-QAM	Dual-Cell	.97	28.0
Dual-Cell HSDPA	Release 8	23	15	64-QAM	Dual-Cell	.82	35.3
Dual-Cell HSDPA	Release 8	24	15	64-QAM	Dual-Cell	.98	42.2
DC-HSDPA w/MIMO	Release 9	25	15	16-QAM	Dual-Cell + MIMO	.81	46.7
DC-HSDPA w/MIMO	Release 9	26	15	16-QAM	Dual-Cell + MIMO	.97	55.9
DC-HSDPA w/MIMO	Release 9	27	15	64-QAM	Dual-Cell + MIMO	.82	70.6
DC-HSDPA w/MIMO	Release 9	28	15	64-QAM	Dual-Cell + MIMO	.98	84.4

16-QAM implies QPSK support, 64-QAM implies 16-QAM and QPSK support. The maximum data rates given in the table are physical layer data rates. Application layer data rate is approximately 85% of that, due to the inclusion of IP headers (overhead information) etc.

Roadmap

The first phase of HSDPA has been specified in the 3rd Generation Partnership Project (3GPP) release 5. Phase one introduces new basic functions and is aimed to achieve peak data rates of 14.0 Mbit/s (see above). Newly introduced are the High Speed Downlink Shared Channels (HS-DSCH), the adaptive modulation QPSK and 16QAM and the High Speed Medium Access protocol (MAC-hs) in base station.

The second phase of HSDPA is specified in the 3GPP release 7 and has been named HSPA Evolved. It can achieve data rates of up to 42 Mbit/s.[1] It introduces antenna array technologies such as beamforming and Multiple-input multiple-output communications (MIMO). Beam forming focuses the transmitted power of an antenna in a beam towards the user's direction. MIMO uses multiple antennas at the sending and receiving side. Deployments are scheduled to begin in the second half of 2008.

Further releases of the standard have introduced dual carrier operation, i.e. the simultaneous use of two 5 MHz carrier. By combining this with MIMO transmission, peak data rates of 84 Mbit/s can be reached under ideal signal conditions.

After HSPA Evolved, the roadmap leads to E-UTRA (Previously "HSOPA"), the technology specified in 3GPP Release 8. This project is called the Long Term Evolution initiative. The first release of LTE offers data rates of over 320 Mbit/s for downlink and over 170 Mbit/s for uplink using OFDMA modulation.[1]

Adoption

As of August 28, 2009, 250 HSDPA networks have commercially launched mobile broadband services in 109 countries. 169 HSDPA networks support 3.6 Mbit/s peak downlink data throughput. A growing number are delivering 21 Mbit/s peak data downlink and 28 Mbit/s. Several others will have this capability by end 2009 and the first 42 Mbit/s network came online in Australia in February 2010. Telstra switches on 42 Mbit/s Next G, plans 84 Mbit/s through the implementation of HSPA+ Dual Carrier plus MIMO technology upgrade in 2011.[4] This protocol is a relatively simple upgrade where UMTS is already deployed.[1] First week in May 2010, Second-ranked Indonesia cellco: Indosat has launched the first DC-HSPA+ 42 Mbit/s fastest commercial network in Asia-Pasific (first operator in Asia and the second in the world after Telstra). Indosat has beaten Australia's Telstra, Singapore's StarHub and Hong Kong's CSL to stake its claim as the first operator in Asia-Pacific to offer theoritical download speeds of 42 Mbit/s via HSPA+.[5] [6]

CDMA2000-EVDO networks had the early lead on performance, and Japanese providers were highly successful benchmarks for it. But lately this seems to be changing in favour of HSDPA as an increasing number of providers worldwide are adopting it. In Australia, Telstra announced that its CDMA-EVDO network would be replaced with a HSDPA network (since named NextG), offering high speed internet, mobile television and traditional telephony and video calling. Rogers Wireless deployed HSDPA system 850/1900 in Canada on April 1, 2007. In July 2008, Bell Canada and Telus announced a joint plan to expand their current shared EVDO/CDMA network to include HSDPA.[7] Bell Canada launched their joint network November 4, 2009, while Telus launched November 5, 2009.[8] In January 2010, T-Mobile USA adopted HSDPA.[9]

Marketing as mobile broadband

During 2007, an increasing number of telcos worldwide began selling HSDPA USB modems as mobile broadband connections. In addition, the popularity of HSDPA landline replacement boxes grew—providing HSDPA for data via Ethernet and WiFi, and ports for connecting traditional landline telephones. Some are marketed with connection speeds of "up to 7.2 Mbit/s",[10] which is only attained under ideal conditions. As a result these services can be slower than expected, especially when in fringe coverage indoors.

See also

- 3GPP Long Term Evolution
- Cellular router
- High-Speed Uplink Packet Access
- High-Speed OFDM Packet Access
- List of device bandwidths
- List of HSDPA mobile phones
- Multi-band
- UMTS
- UMTS frequency bands

References

[1] HSPA mobile broadband today (http://www.gsmworld.com/HSPA)
[2] 3GPP TS 25.306 v9.0.0 http://www.3gpp.org/ftp/Specs/html-info/25306.htm
[3] The maximal code rate is not limited. A value close to 1 in this column indicates that the maximum data rate can be achieved only in ideal conditions. The device is therefore connected directly to the transmitter to demonstrate these data rates.
[4] Telstra switches on 42 Mbps Next G, plans 84 Mbps upgrade in 2011 | Comms Day http://www.commsday.com/commsday/?p=789
[5] Indosat first in Asia to launch 42 Mbps HSPA+ http://www.telecomasia.net/print/17244
[6] Indosat gears up for 4G and launches Asia's fastest network - Ericsson http://www.ericsson.com/news/142992
[7] "Telus, Bell Announce Switch from CDMA to HSDPA" (http://www.iphonealley.com/news/telus-bell-announce-switch-from-cdma-to-hsdpa). .
[8] Marlow, Iain (3 November 2009). "Bell, Telus launch high-speed networks" (http://www.thestar.com/business/article/720096--bell-races-to-launch-high-speed-network-one-day-ahead-of-telus). *Toronto Star*. .
[9] http://www.pcworld.com/businesscenter/article/185916/tmobile_usa_finishes_upgrade_to_hspa_72.html
[10] Vodafone UK 7.2 MBs service (http://online.vodafone.co.uk/dispatch/Portal/appmanager/vodafone/wrp?_nfpb=true&_pageLabel=template12&pageID=PPP_0026)

Further reading

- Sauter, Martin (2006). *Communication Systems for the Mobile Information Society*. Chichester: John Wiley. ISBN 0470026766.

External links

- GSM Association on HSPA (http://www.gsmworld.com/HSPA)
- Understand HSDPA's implementation challenges (http://www.mobilehandsetdesignline.com/howto/170703686)

Sony Ericsson XPERIA X10 Mini Pro

The **Sony Ericsson Xperia X10 Mini Pro** is a mobile telephone released by Sony Ericsson on the 24 May 2010.

The device is an upgrade of the similar X10 Mini with many of the internal specifications being identical. The major differences between the X10 Mini and X10 Mini Pro are a slide-out full QWERTY Keyboard, and the Pro having slightly larger dimensions (3.5 × 2.0 × 0.7 inches opposed to 3.3 × 2.0 × 0.6 inches).

The X10 Mini and X10 Mini Pro are designed to look similar and share functionality with the larger Xperia X10, but are internally very different devices. The X10 Mini and X10 Mini Pro lack Sony Ericsson's "Mediascape" media-management software, but include "Timescape" as well as the proprietary "Rachael" UI.

The X10 Mini Pro (as well as the X10 and X10 Mini) runs on Android 1.6, with an update to 2.1 being rolled out from Sunday 31st October 2010, to Tuesday 30th November 2010.

References

- Sony Ericsson X10 Mini Official Page [1]
- Sony Ericsson X10 Mini Pro Official Page [2]

Bluetooth

Bluetooth logo

Bluetooth is an open wireless technology standard for exchanging data over short distances (using short wavelength radio transmissions) from fixed and mobile devices, creating personal area networks (PANs) with high levels of security. Created by telecoms vendor Ericsson in 1994,[1] it was originally conceived as a wireless alternative to RS-232 data cables. It can connect several devices, overcoming problems of synchronization. Today Bluetooth is managed by the Bluetooth Special Interest Group.

Name and logo

The word Bluetooth is an anglicised version of the Scandinavian *Blåtand*, the epithet of the tenth-century king Harald I of Denmark and parts of Norway who united dissonant Danish tribes into a single kingdom. The implication is that Bluetooth does the same with communications protocols, uniting them into one universal standard.[2] [3] [4]

The Bluetooth logo is a bind rune merging the Younger Futhark runes ᚼ (Hagall) ( ) and ᛒ (Bjarkan) ( ), Harald's initials.

Implementation

Bluetooth uses a radio technology called frequency-hopping spread spectrum, which chops up the data being sent and transmits chunks of it on up to 79 bands (1 MHz each) in the range 2402-2480 MHz. This range is in the globally unlicensed Industrial, Scientific and Medical (ISM) 2.4 GHz short-range radio frequency band.

Originally Gaussian frequency-shift keying (GFSK) modulation was the only modulation scheme available; subsequently, since the introduction of Bluetooth 2.0+EDR, π/4-DQPSK and 8DPSK modulation may also be used between compatible devices. Devices functioning with GFSK are said to be operating in basic rate (BR) mode where a gross data rate of 1 Mbit/s is possible. The term enhanced data rate (EDR) is used to describe π/4-DPSK and 8DPSK schemes, each giving 2 and 3 Mbit/s respectively. The combination of these (BR and EDR) modes in Bluetooth radio technology is classified as a "BR/EDR radio".

Bluetooth is a packet-based protocol with a master-slave structure. One master may communicate with up to 7 slaves in a piconet; all devices share the master's clock. Packet exchange is based on the basic clock, defined by the master, which ticks at 312.5 µs intervals. Two clock ticks make up a slot of 625 µs; two slots make up a slot pair of 1250 µs. In the simple case of single-slot packets the master transmits in even slots and receives in odd slots; the slave, conversely, receives in even slots and transmits in odd slots. Packets may be 1, 3 or 5 slots long but in all cases the master transmit will begin in even slots and the slave transmit in odd slots.

Bluetooth provides a secure way to connect and exchange information between devices such as faxes, mobile phones, telephones, laptops, personal computers, printers, Global Positioning System (GPS) receivers, digital cameras, and video game consoles.

The Bluetooth specifications are developed and licensed by the Bluetooth Special Interest Group (SIG). The Bluetooth SIG consists of more than 13,000 companies in the areas of telecommunication, computing, networking, and consumer electronics.[5]

To be marketed as a Bluetooth device, it must be qualified to standards defined by the SIG.

Communication and connection

A master Bluetooth device can communicate with up to seven devices in a piconet. The devices can switch roles, by agreement, and the slave can become the master at any time.

At any given time, data can be transferred between the master and one other device.

The master chooses which slave device to address; typically, it switches rapidly from one device to another in a round-robin fashion. Simultaneous transmission from the master to multiple other devices is possible via broadcast mode, but this capability is infrequently used in practice.

The Bluetooth Core Specification provides for the connection of two or more piconets to form a scatternet, in which certain devices serve as bridges, simultaneously playing the master role in one piconet and the slave role in another.

Many USB Bluetooth adapters or "dongles" are available, some of which also include an IrDA adapter. Older (pre-2003) Bluetooth dongles, however, have limited capabilities, offering only the Bluetooth Enumerator and a less-powerful Bluetooth Radio incarnation. Such devices can link computers with Bluetooth with a distance of 100 meters, but they do not offer much in the way of services that modern adapters do.

Uses

Bluetooth is a standard communications protocol primarily designed for low power consumption, with a short range (power-class-dependent: 100 m, 10 m and 1 m, but ranges vary in practice; see table below) based on low-cost transceiver microchips in each device.[6] Because the devices use a radio (broadcast) communications system, they do not have to be in line of sight of each other.[5]

Class	Maximum Permitted Power		Range (approximate)
	mW	dBm	
Class 1	100	20	~100 meters
Class 2	2.5	4	~10 meters
Class 3	1	0	~1 meters

In most cases the effective range of class 2 devices is extended if they connect to a class 1 transceiver, compared to a pure class 2 network. This is accomplished by the higher sensitivity and transmission power of Class 1 devices.[7]

Version	Data Rate
Version 1.2	1 Mbit/s
Version 2.0 + EDR	3 Mbit/s
Version 3.0 + HS	24 Mbit/s

While the Bluetooth Core Specification does mandate minimums for range, the range of the technology is application specific and is not limited. Manufacturers may tune their implementations to the range needed to support individual use cases.

Bluetooth profiles

To use Bluetooth wireless technology, a device must be able to interpret certain Bluetooth profiles, which are definitions of possible applications and specify general behaviors that Bluetooth enabled devices use to communicate with other Bluetooth devices. There are a wide range of Bluetooth profiles that describe many different types of applications or use cases for devices.[8]

List of applications

A typical Bluetooth mobile phone headset.

- Wireless control of and communication between a mobile phone and a handsfree headset. This was one of the earliest applications to become popular.
- Wireless networking between PCs in a confined space and where little bandwidth is required.
- Wireless communication with PC input and output devices, the most common being the mouse, keyboard and printer.
- Transfer of files, contact details, calendar appointments, and reminders between devices with OBEX.
- Replacement of traditional wired serial communications in test equipment, GPS receivers, medical equipment, bar code scanners, and traffic control devices.
- For controls where infrared was traditionally used.
- For low bandwidth applications where higher USB bandwidth is not required and cable-free connection desired.
- Sending small advertisements from Bluetooth-enabled advertising hoardings to other, discoverable, Bluetooth devices.[9]
- Wireless bridge between two Industrial Ethernet (e.g., PROFINET) networks.
- Three seventh-generation game consoles, Nintendo's Wii[10] and Sony's PlayStation 3 and PSP Go, use Bluetooth for their respective wireless controllers.
- Dial-up internet access on personal computers or PDAs using a data-capable mobile phone as a wireless modem like Novatel mifi.
- Short range transmission of health sensor data from medical devices to mobile phone, set-top box or dedicated telehealth devices.[11]
- Allowing a DECT phone to ring and answer calls on behalf of a nearby cell phone
- Real-time location systems (RTLS), are used to track and identify the location of objects in real-time using “Nodes” or “tags” attached to, or embedded in the objects tracked, and “Readers” that receive and process the wireless signals from these tags to determine their locations[12]

Bluetooth vs. Wi-Fi IEEE 802.11 in networking

Bluetooth and Wi-Fi have many applications: setting up networks, printing, or transferring files.

Wi-Fi is intended for resident equipment and its applications. The category of applications is outlined as WLAN, the wireless local area networks. Wi-Fi is intended as a replacement for cabling for general local area network access in work areas.

Bluetooth is intended for non-resident equipment and its applications. The category of applications is outlined as the wireless personal area network (WPAN). Bluetooth is a replacement for cabling in a variety of personally carried applications in any ambiance and can also support fixed location applications such as smart energy functionality in the home (thermostats, etc.).

Wi-Fi is wireless version of a traditional Ethernet network, and requires configuration to set up shared resources, transmit files, and to set up audio links (for example, headsets and hands-free devices). Wi-Fi uses the same radio frequencies as Bluetooth, but with higher power, resulting in a faster connection and better range from the base station. The nearest equivalents in Bluetooth are the DUN profile, which allows devices to act as modem interfaces, and the PAN profile, which allows for ad-hoc networking.

Bluetooth devices

A Bluetooth USB dongle with a 100 m range. The MacBook Pro, shown, also has a built in Bluetooth adaptor.

Bluetooth exists in many products, such as the iPod Touch, Lego Mindstorms NXT, PlayStation 3, PSP Go, telephones, the Wii, and some high definition headsets, modems, and watches. "Watch" [13]. Bluetooth.com. Retrieved 2010-09-04. The technology is useful when transferring information between two or more devices that are near each other in low-bandwidth situations. Bluetooth is commonly used to transfer sound data with telephones (i.e., with a Bluetooth headset) or byte data with hand-held computers (transferring files).

Bluetooth protocols simplify the discovery and setup of services between devices. Bluetooth devices can advertise all of the services they provide. [14] This makes using services easier because more of the security, network address and permission configuration can be automated than with many other network types.

Computer requirements

A typical Bluetooth USB dongle.

A personal computer that does not have embedded Bluetooth can be used with a Bluetooth adapter or "dongle" that will enable the PC to communicate with other Bluetooth devices (such as mobile phones, mice and keyboards). While some desktop computers and most recent laptops come with a built-in Bluetooth radio, others will require an external one in the form of a dongle.

Unlike its predecessor, IrDA, which requires a separate adapter for each device, Bluetooth allows multiple devices to communicate with a computer over a single adapter.

Operating system support

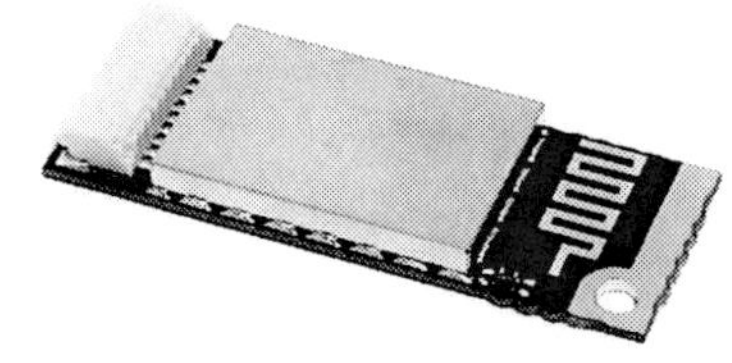
An internal notebook Bluetooth card (14×36×4 mm).

Apple has supported Bluetooth since Mac OS X v10.2 which was released in 2002.[15]

For Microsoft platforms, Windows XP Service Pack 2 and SP3 releases have native support for Bluetooth 1.1, 2.0 and 2.0+EDR.[16] Previous versions required users to install their Bluetooth adapter's own drivers, which were not directly supported by Microsoft.[17] Microsoft's own Bluetooth dongles (packaged with their Bluetooth computer devices) have no external drivers and thus require at least Windows XP Service Pack 2. Windows Vista RTM/SP1 with the Feature Pack for Wireless or Windows Vista SP2 support Bluetooth 2.1+EDR.[16] Windows 7 supports Bluetooth 2.1+EDR and Extended Inquiry Response (EIR).[16]

The Windows XP and Windows Vista/Windows 7 Bluetooth stacks support the following Bluetooth profiles natively: PAN, SPP, DUN, HID, HCRP. The Windows XP stack can be replaced by a third party stack which may support more profiles or newer versions of Bluetooth. The Windows Vista/Windows 7 Bluetooth stack supports vendor-supplied additional profiles without requiring the Microsoft stack to be replaced.[16]

Linux has two popular Bluetooth stacks, BlueZ and Affix. The BlueZ stack is included with most Linux kernels and was originally developed by Qualcomm.[18] The Affix stack was developed by Nokia. FreeBSD features Bluetooth support since its 5.0 release. NetBSD features Bluetooth support since its 4.0 release. Its Bluetooth stack has been ported to OpenBSD as well.

Mobile phone requirements

A Bluetooth-enabled mobile phone is able to pair with many devices. To ensure the broadest support of feature functionality together with legacy device support, the Open Mobile Terminal Platform (OMTP) forum has published a recommendations paper, entitled "Bluetooth Local Connectivity".[19]

Specifications and features

The Bluetooth specification was developed in 1994 by Jaap Haartsen and Sven Mattisson, who were working for Ericsson in Lund, Sweden.[20] The specification is based on frequency-hopping spread spectrum technology.

The specifications were formalized by the Bluetooth Special Interest Group (SIG). The SIG was formally announced on May 20, 1998. Today it has a membership of over 13,000 companies worldwide. It was established by Ericsson, IBM, Intel, Toshiba, Motorola and Nokia, and later joined by many other companies.

Bluetooth v1.0 and v1.0B

Versions 1.0 and 1.0B had many problems, and manufacturers had difficulty making their products interoperable. Versions 1.0 and 1.0B also included mandatory Bluetooth hardware device address (BD_ADDR) transmission in the Connecting process (rendering anonymity impossible at the protocol level), which was a major setback for certain services planned for use in Bluetooth environments.

Bluetooth v1.1

- Ratified as IEEE Standard 802.15.1-2002[21]
- Many errors found in the 1.0B specifications were fixed.
- Added support for non-encrypted channels.
- Received Signal Strength Indicator (RSSI).

Bluetooth v1.2

This version is backward compatible with 1.1 and the major enhancements include the following:

- Faster Connection and Discovery
- *Adaptive frequency-hopping spread spectrum (AFH)*, which improves resistance to radio frequency interference by avoiding the use of crowded frequencies in the hopping sequence.
- Higher transmission speeds in practice, up to 721 kbit/s,[22] than in 1.1.
- Extended Synchronous Connections (eSCO), which improve voice quality of audio links by allowing retransmissions of corrupted packets, and may optionally increase audio latency to provide better support for concurrent data transfer.
- Host Controller Interface (HCI) support for three-wire UART.
- Ratified as IEEE Standard 802.15.1-2005[23]
- Introduced Flow Control and Retransmission Modes for L2CAP.

Bluetooth v2.0 + EDR

This version of the Bluetooth Core Specification was released in 2004 and is backward compatible with the previous version 1.2. The main difference is the introduction of an Enhanced Data Rate (EDR) for faster data transfer. The nominal rate of EDR is about 3 Mbit/s, although the practical data transfer rate is 2.1 Mbit/s.[22] EDR uses a combination of GFSK and Phase Shift Keying modulation (PSK) with two variants, π/4-DQPSK and 8DPSK.[24] EDR can provide a lower power consumption through a reduced duty cycle.

The specification is published as "Bluetooth v2.0 + EDR" which implies that EDR is an optional feature. Aside from EDR, there are other minor improvements to the 2.0 specification, and products may claim compliance to "Bluetooth v2.0" without supporting the higher data rate. At least one commercial device states "Bluetooth v2.0 without EDR" on its data sheet.[25]

Bluetooth v2.1 + EDR

Bluetooth Core Specification Version 2.1 + EDR is fully backward compatible with 1.2, and was adopted by the Bluetooth SIG on July 26, 2007.[24]

The headline feature of 2.1 is secure simple pairing (SSP): this improves the pairing experience for Bluetooth devices, while increasing the use and strength of security. See the section on Pairing below for more details.[26]

2.1 allows various other improvements, including "Extended inquiry response" (EIR), which provides more information during the inquiry procedure to allow better filtering of devices before connection; sniff subrating, which reduces the power consumption in low-power mode

Bluetooth v3.0 + HS

Version 3.0 + HS of the Bluetooth Core Specification[24] was adopted by the Bluetooth SIG [27] on April 21, 2009. Bluetooth 3.0+HS supports theoretical data transfer speeds of up to 24 Mbit/s, though not over the Bluetooth link itself. Instead, the Bluetooth link is used for negotiation and establishment, and the high data rate traffic is carried over a colocated 802.11 link. Its main new feature is AMP (Alternate MAC/PHY), the addition of 802.11 as a high speed transport. Two technologies had been anticipated for AMP: 802.11 and UWB, but UWB is missing from the specification.[28]

The High-Speed part of the specification is not mandatory, and hence only devices sporting the "+HS" will actually support the Bluetooth over Wifi high-speed data transfer. A Bluetooth 3.0 device without the HS suffix will not support High Speed, and needs to only support Unicast Connectionless Data (UCD), as shown in the Bluetooth 3.0+HS specification [29], Vol0, section 4.1 Specification Naming Conventions.

Alternate MAC/PHY

Enables the use of alternative MAC and PHYs for transporting Bluetooth profile data. The Bluetooth radio is still used for device discovery, initial connection and profile configuration, however when large quantities of data need to be sent, the high speed alternate MAC PHY 802.11 (typically associated with Wi-Fi) will be used to transport the data. This means that the proven low power connection models of Bluetooth are used when the system is idle, and the low power per bit radios are used when large quantities of data need to be sent.

Unicast connectionless data

Permits service data to be sent without establishing an explicit L2CAP channel. It is intended for use by applications that require low latency between user action and reconnection/transmission of data. This is only appropriate for small amounts of data.

Enhanced Power Control

Updates the power control feature to remove the open loop power control, and also to clarify ambiguities in power control introduced by the new modulation schemes added for EDR. Enhanced power control removes the ambiguities by specifying the behaviour that is expected. The feature also adds closed loop power control, meaning RSSI filtering can start as the response is received. Additionally, a "go straight to maximum power" request has been introduced. This is expected to deal with the headset link loss issue typically observed when a user puts their phone into a pocket on the opposite side to the headset.

Bluetooth v4.0

On June 12, 2007, Nokia and Bluetooth SIG had announced that Wibree will be a part of the Bluetooth specification, as an ultra-low power Bluetooth technology.[30]

On December 17, 2009, the Bluetooth SIG adopted Bluetooth low energy technology as the hallmark feature of the version 4.0.[31] The provisional names *Wibree* and *Bluetooth ULP* (Ultra Low Power) are abandoned.

On April 21, 2010, the Bluetooth SIG completed the Bluetooth Core Specification version 4.0, which includes *Classic Bluetooth*, *Bluetooth high speed* and *Bluetooth low energy* protocols. Bluetooth high speed is based on Wi-Fi, and Classic Bluetooth consists of legacy Bluetooth protocols.

Bluetooth low energy

Bluetooth low energy is an enhancement to the Bluetooth standard that was introduced in Bluetooth v4.0. It allows two types of implementation, dual-mode and single-mode. In a dual-mode implementation, Bluetooth low energy functionality is integrated into an existing Classic Bluetooth controller. The resulting architecture shares much of Classic Bluetooth's existing radio and functionality resulting in a minimal cost increase compared to Classic Bluetooth. Additionally, manufacturers can use current Classic Bluetooth (Bluetooth v2.1 + EDR or Bluetooth v3.0 + HS) chips with the new low energy stack, enhancing the development of Classic Bluetooth enabled devices with new capabilities.

Single-mode chips, which will enable highly integrated and compact devices, will feature a lightweight Link Layer providing ultra-low power idle mode operation, simple device discovery, and reliable point-to-multipoint data transfer with advanced power-save and secure encrypted connections at the lowest possible cost. The Link Layer in these controllers will enable Internet connected sensors to schedule Bluetooth low energy traffic between Bluetooth transmissions.

Expected use cases for Bluetooth low energy technology include sports and fitness, security and proximity and smart energy. Bluetooth low energy technology is designed for devices to have a battery life of up to one year such as those powered by coin-cell batteries. These types of devices include watches that will use Bluetooth low energy technology to display Caller ID information and sports sensors that will be used to monitor the wearer's heart rate during exercise. The Medical Devices Working Group of the Bluetooth SIG is also creating a medical devices profile and associated protocols to enable Bluetooth applications for this vertical market.

Future

Broadcast channel

> Enables Bluetooth information points. This will drive the adoption of Bluetooth into mobile phones, and enable advertising models based on users pulling information from the information points, and not based on the object push model that is used in a limited way today.

Topology management

> Enables the automatic configuration of the piconet topologies especially in scatternet situations that are becoming more common today. This should all be invisible to users of the technology, while also making the technology "just work."

QoS improvements

> Enable audio and video data to be transmitted at a higher quality, especially when best effort traffic is being transmitted in the same piconet.

UWB for AMP

The high speed (AMP) feature of Bluetooth v3.0 is based on 802.11, but the AMP mechanism was designed to be usable with other radios as well. It was originally intended for UWB, but the WiMedia Alliance, the body responsible for the flavor of UWB intended for Bluetooth, announced in March 2009 that it was disbanding.

On March 16, 2009, the WiMedia Alliance announced it was entering into technology transfer agreements for the WiMedia Ultra-wideband (UWB) specifications. WiMedia has transferred all current and future specifications, including work on future high speed and power optimized implementations, to the Bluetooth Special Interest Group (SIG), Wireless USB Promoter Group and the USB Implementers Forum. After the successful completion of the technology transfer, marketing and related administrative items, the WiMedia Alliance will cease operations.[32] [33] [34] [35] [36] [37]

In October 2009 the Bluetooth Special Interest Group suspended development of UWB as part of the alternative MAC/PHY, Bluetooth v3.0 + HS solution. A small, but significant, number of former WiMedia members had not and would not sign up to the necessary agreements for the IP transfer. The Bluetooth SIG is now in the process of evaluating other options for its longer term roadmap.[38]

Technical information

Bluetooth protocol stack

"Bluetooth is defined as a layer protocol architecture consisting of core protocols, cable replacement protocols, telephony control protocols, and adopted protocols."[39] Mandatory protocols for all Bluetooth stacks are: LMP, L2CAP and SDP. Additionally, these protocols are almost universally supported: HCI and RFCOMM.

LMP (Link Management Protocol)

Used for control of the radio link between two devices. Implemented on the controller.

L2CAP (Logical Link Control & Adaptation Protocol)

Used to multiplex multiple logical connections between two devices using different higher level protocols. Provides segmentation and reassembly of on-air packets.

In Basic mode, L2CAP provides packets with a payload configurable up to 64kB, with 672 bytes as the default MTU, and 48 bytes as the minimum mandatory supported MTU.

In Retransmission & Flow Control modes, L2CAP can be configured for reliable or isochronous data per channel by performing retransmissions and CRC checks.

Bluetooth Core Specification Addendum 1 adds two additional L2CAP modes to the core specification. These modes effectively deprecate original Retransmission and Flow Control modes:

- **Enhanced Retransmission Mode** (ERTM): This mode is an improved version of the original retransmission mode. This mode provides a reliable L2CAP channel.
- **Streaming Mode** (SM): This is a very simple mode, with no retransmission or flow control. This mode provides an unreliable L2CAP channel.

Reliability in any of these modes is optionally and/or additionally guaranteed by the lower layer Bluetooth BDR/EDR air interface by configuring the number of retransmissions and flush timeout (time after which the radio will flush packets). In-order sequencing is guaranteed by the lower layer.

Only L2CAP channels configured in ERTM or SM may be operated over AMP logical links.

SDP (Service Discovery Protocol)

Service Discovery Protocol (SDP) allows a device to discover services supported by other devices, and their associated parameters. For example, when connecting a mobile phone to a Bluetooth headset, SDP will be used for determining which Bluetooth profiles are supported by the headset (Headset Profile, Hands Free Profile, Advanced Audio Distribution Profile (A2DP) etc.) and the protocol multiplexer settings needed to connect to each of them. Each service is identified by a Universally Unique Identifier (UUID), with official services (Bluetooth profiles) assigned a short form UUID (16 bits rather than the full 128)

HCI (Host/Controller Interface)

Standardised communication between the host stack (e.g., a PC or mobile phone OS) and the controller (the Bluetooth IC). This standard allows the host stack or controller IC to be swapped with minimal adaptation.

There are several HCI transport layer standards, each using a different hardware interface to transfer the same command, event and data packets. The most commonly used are USB (in PCs) and UART (in mobile phones and PDAs).

In Bluetooth devices with simple functionality (e.g., headsets) the host stack and controller can be implemented on the same microprocessor. In this case the HCI is optional, although often implemented as an internal software interface.

RFCOMM (Serial Port Emulation)

Radio frequency communications (RFCOMM) is a cable replacement protocol used to create a virtual serial data stream. RFCOMM provides for binary data transport and emulates EIA-232 (formerly RS-232) control signals over the Bluetooth baseband layer.

RFCOMM provides a simple reliable data stream to the user, similar to TCP. It is used directly by many telephony related profiles as a carrier for AT commands, as well as being a transport layer for OBEX over Bluetooth.

Many Bluetooth applications use RFCOMM because of its widespread support and publicly available API on most operating systems. Additionally, applications that used a serial port to communicate can be quickly ported to use RFCOMM.

BNEP (Bluetooth Network Encapsulation Protocol)

BNEP is used for transferring another protocol stack's data via an L2CAP channel. It's main purpose is the transmission of IP packets in the Personal Area Networking Profile. BNEP performs a similar function to SNAP in Wireless LAN.

AVCTP (Audio/Video Control Transport Protocol)

Used by the remote control profile to transfer AV/C commands over an L2CAP channel. The music control buttons on a stereo headset use this protocol to control the music player.

AVDTP (Audio/Video Distribution Transport Protocol)

Used by the advanced audio distribution profile (A2DP) to stream music to stereo headsets over an L2CAP channel. Intended to be used by video distribution profile.

Telephony control protocol

Telephony control protocol-binary (TCS BIN) is the bit-oriented protocol that defines the call control signaling for the establishment of voice and data calls between Bluetooth devices. Additionally, "TCS BIN defines mobility management procedures for handling groups of Bluetooth TCS devices."

TCS-BIN is only used by the cordless telephony profile, which failed to attract implementers. As such it is only of historical interest.

Adopted protocols

Adopted protocols are defined by other standards-making organizations and incorporated into Bluetooth's protocol stack, allowing Bluetooth to create protocols only when necessary. The adopted protocols include:

Point-to-Point Protocol (PPP)

Internet standard protocol for transporting IP datagrams over a point-to-point link.

TCP/IP/UDP

Foundation Protocols for TCP/IP protocol suite

Object Exchange Protocol (OBEX)

Session-layer protocol for the exchange of objects, providing a model for object and operation representation

Wireless Application Environment/Wireless Application Protocol (WAE/WAP)

WAE specifies an application framework for wireless devices and WAP is an open standard to provide mobile users access to telephony and information services.[39]

Baseband Error Correction

Three types of error correction are implemented in Bluetooth systems,

- 1/3 rate forward error correction (FEC)
- 2/3 rate FEC
- Automatic repeat-request (ARQ)

Setting up connections

Any Bluetooth device in *discoverable mode* will transmit the following information on demand:

- Device name
- Device class
- List of services
- Technical information (for example: device features, manufacturer, Bluetooth specification used, clock offset)

Any device may perform an inquiry to find other devices to connect to, and any device can be configured to respond to such inquiries. However, if the device trying to connect knows the address of the device, it always responds to direct connection requests and transmits the information shown in the list above if requested. Use of a device's services may require pairing or acceptance by its owner, but the connection itself can be initiated by any device and held until it goes out of range. Some devices can be connected to only one device at a time, and connecting to them prevents them from connecting to other devices and appearing in inquiries until they disconnect from the other device.

Every device has a unique 48-bit address. However, these addresses are generally not shown in inquiries. Instead, friendly Bluetooth names are used, which can be set by the user. This name appears when another user scans for devices and in lists of paired devices.

Most phones have the Bluetooth name set to the manufacturer and model of the phone by default. Most phones and laptops show only the Bluetooth names and special programs are required to get additional information about remote devices. This can be confusing as, for example, there could be several phones in range named T610 (see Bluejacking).

Pairing

Motivation

Many of the services offered over Bluetooth can expose private data or allow the connecting party to control the Bluetooth device. For security reasons it is therefore necessary to control which devices are allowed to connect to a given Bluetooth device. At the same time, it is useful for Bluetooth devices to automatically establish a connection without user intervention as soon as they are in range.

To resolve this conflict, Bluetooth uses a process called *pairing*, which is generally manually started by a device user—making that device's Bluetooth link visible to other devices. Two devices need to be *paired* to communicate with each other; the pairing process is typically triggered automatically the first time a device receives a connection request from a device with which it is not yet paired. Once a pairing has been established it is remembered by the devices, which can then connect to each without user intervention. When desired, the pairing relationship can later be removed by the user.

Implementation

During the pairing process, the two devices involved establish a relationship by creating a shared secret known as a *link key*. If a link key is stored by both devices they are said to be *paired* or *bonded*. A device that wants to communicate only with a bonded device can cryptographically authenticate the identity of the other device, and so be sure that it is the same device it previously paired with. Once a link key has been generated, an authenticated ACL link between the devices may be encrypted so that the data that they exchange over the airwaves is protected against eavesdropping.

Link keys can be deleted at any time by either device. If done by either device this will implicitly remove the bonding between the devices; so it is possible for one of the devices to have a link key stored but not be aware that it is no longer bonded to the device associated with the given link key.

Bluetooth services generally require either encryption or authentication, and as such require pairing before they allow a remote device to use the given service. Some services, such as the Object Push Profile, elect not to explicitly require authentication or encryption so that pairing does not interfere with the user experience associated with the service use-cases.

Pairing mechanisms

Pairing mechanisms have changed significantly with the introduction of Secure Simple Pairing in Bluetooth v2.1. The following summarizes the pairing mechanisms:

- **Legacy pairing**: This is the only method available in Bluetooth v2.0 and before. Each device must enter a PIN code; pairing is only successful if both devices enter the same PIN code. Any 16-byte UTF-8 string may be used as a PIN code, however not all devices may be capable of entering all possible PIN codes.
  - **Limited input devices**: The obvious example of this class of device is a Bluetooth Hands-free headset, which generally have few inputs. These devices usually have a *fixed PIN*, for example "0000" or "1234", that are hard-coded into the device.
  - **Numeric input devices**: Mobile phones are classic examples of these devices. They allow a user to enter a numeric value up to 16 digits in length.
  - **Alpha-numeric input devices**: PCs and smartphones are examples of these devices. They allow a user to enter full UTF-8 text as a PIN code. If pairing with a less capable device the user needs to be aware of the input limitations on the other device, there is no mechanism available for a capable device to determine how it

should limit the available input a user may use.

- **Secure Simple Pairing** (SSP): This is required by Bluetooth v2.1. A Bluetooth v2.1 device may only use legacy pairing to interoperate with a v2.0 or earlier device. Secure Simple Pairing uses a form of public key cryptography, and has the following modes of operation:
  - **Just works**: As implied by the name, this method just works. No user interaction is required; however, a device may prompt the user to confirm the pairing process. This method is typically used by headsets with very limited IO capabilities, and is more secure than the fixed PIN mechanism which is typically used for legacy pairing by this set of limited devices. This method provides no man in the middle (MITM) protection.
  - **Numeric comparison**: If both devices have a display and at least one can accept a binary Yes/No user input, they may use Numeric Comparison. This method displays a 6-digit numeric code on each device. The user should compare the numbers to ensure they are identical. If the comparison succeeds, the user(s) should confirm pairing on the device(s) that can accept an input. This method provides MITM protection, assuming the user confirms on both devices and actually performs the comparison properly.
  - **Passkey Entry**: This method may be used between a device with a display and a device with numeric keypad entry (such as a keyboard), or two devices with numeric keypad entry. In the first case, the display is used to show a 6-digit numeric code to the user, who then enters the code on the keypad. In the second case, the user of each device enters the same 6-digit number. Both cases provide MITM protection.
  - **Out of band** (OOB): This method uses an external means of communication (such as NFC) to exchange some information used in the pairing process. Pairing is completed using the Bluetooth radio, but requires information from the OOB mechanism. This provides only the level of MITM protection that is present in the OOB mechanism.

SSP is considered simple for the following reasons:

- In most cases, it does not require a user to generate a passkey.
- For use-cases not requiring MITM protection, user interaction has been eliminated.
- For *numeric comparison*, MITM protection can be achieved with a simple equality comparison by the user.
- Using OOB with NFC will enable pairing when devices simply get close, rather than requiring a lengthy discovery process.

Security Concerns

Prior to Bluetooth v2.1, encryption is not required and can be turned off at any time. Moreover, the encryption key is only good for approximately 23.5 hours; using a single encryption key longer than this time allows simple XOR attacks to retrieve the encryption key.

- Turning off encryption is required for several normal operations, so it is problematic to detect if encryption is disabled for a valid reason or for a security attack.
- Bluetooth v2.1 addresses this in the following ways:
  - Encryption is required for all non-SDP (Service Discovery Protocol) connections
  - A new Encryption Pause and Resume feature is used for all normal operations requiring encryption to be disabled. This enables easy identification of normal operation from security attacks.
  - The encryption key is required to be refreshed before it expires.

Link keys may be stored on the device file system, not on the Bluetooth chip itself. Many Bluetooth chip manufacturers allow link keys to be stored on the device; however, if the device is removable this means that the link key will move with the device.

Air interface

The protocol operates in the license-free ISM band at 2.402-2.480 GHz.[40] To avoid interfering with other protocols that use the 2.45 GHz band, the Bluetooth protocol divides the band into 79 channels (each 1 MHz wide) and changes channels up to 1600 times per second. Implementations with versions 1.1 and 1.2 reach speeds of 723.1 kbit/s. Version 2.0 implementations feature Bluetooth Enhanced Data Rate (EDR) and reach 2.1 Mbit/s. Technically, version 2.0 devices have a higher power consumption, but the three times faster rate reduces the transmission times, effectively reducing power consumption to half that of 1.x devices (assuming equal traffic load).

Security

Overview

Bluetooth implements confidentiality, authentication and key derivation with custom algorithms based on the SAFER+ block cipher. Bluetooth key generation is generally based on a Bluetooth PIN, which must be entered into both devices. This procedure might be modified if one of the devices has a fixed PIN (e.g., for headsets or similar devices with a restricted user interface). During pairing, an initialization key or master key is generated, using the E22 algorithm.[41] The E0 stream cipher is used for encrypting packets, granting confidentiality and is based on a shared cryptographic secret, namely a previously generated link key or master key. Those keys, used for subsequent encryption of data sent via the air interface, rely on the Bluetooth PIN, which has been entered into one or both devices.

An overview of Bluetooth vulnerabilities exploits was published in 2007 by Andreas Becker.[42]

In September 2008, the National Institute of Standards and Technology (NIST) published a Guide to Bluetooth Security that will serve as reference to organizations on the security capabilities of Bluetooth and steps for securing Bluetooth technologies effectively. While Bluetooth has its benefits, it is susceptible to denial of service attacks, eavesdropping, man-in-the-middle attacks, message modification, and resource misappropriation. Users/organizations must evaluate their acceptable level of risk and incorporate security into the lifecycle of Bluetooth devices. To help mitigate risks, included in the NIST document are security checklists with guidelines and recommendations for creating and maintaining secure Bluetooth piconets, headsets, and smart card readers.[43]

Bluetooth v2.1 - finalized in 2007 with consumer devices first appearing in 2009 - makes significant changes to Bluetooth's security, including pairing. See the #Pairing mechanisms section for more about these changes.

Bluejacking

Bluejacking is the sending of either a picture or a message from one user to an unsuspecting user through *Bluetooth* wireless technology. Common applications include short messages (e.g., "You've just been bluejacked!"). [44] Bluejacking does not involve the removal or alteration of any data from the device. Bluejacking can also involve taking control of a mobile wirelessly and phoning a premium rate line, owned by the bluejacker.

History of security concerns

Early

In 2001, Jakobsson and Wetzel from Bell Laboratories discovered flaws in the Bluetooth pairing protocol and also pointed to vulnerabilities in the encryption scheme.[45] In 2003, Ben and Adam Laurie from A.L. Digital Ltd. discovered that serious flaws in some poor implementations of Bluetooth security may lead to disclosure of personal data.[46] In a subsequent experiment, Martin Herfurt from the trifinite.group was able to do a field-trial at the CeBIT fairgrounds, showing the importance of the problem to the world. A new attack called BlueBug was used for this experiment.[47] In 2004 the first purported virus using Bluetooth to spread itself among mobile phones appeared on the Symbian OS.[48] The virus was first described by Kaspersky Lab and requires users to confirm the installation of unknown software before it can propagate. The virus was written as a proof-of-concept by a group of virus writers known as "29A" and sent to anti-virus groups. Thus, it should be regarded as a potential (but not real) security threat to Bluetooth technology or Symbian OS since the virus has never spread outside of this system. In August 2004, a world-record-setting experiment (see also Bluetooth sniping) showed that the range of Class 2 Bluetooth radios could be extended to 1.78 km (1.08 mile) with directional antennas and signal amplifiers.[49] This poses a potential security threat because it enables attackers to access vulnerable Bluetooth devices from a distance beyond expectation. The attacker must also be able to receive information from the victim to set up a connection. No attack can be made against a Bluetooth device unless the attacker knows its Bluetooth address and which channels to transmit on.

2005

In January 2005, a mobile malware worm known as Lasco.A began targeting mobile phones using Symbian OS (Series 60 platform) using Bluetooth enabled devices to replicate itself and spread to other devices. The worm is self-installing and begins once the mobile user approves the transfer of the file (velasco.sis) from another device. Once installed, the worm begins looking for other Bluetooth enabled devices to infect. Additionally, the worm infects other .SIS files on the device, allowing replication to another device through use of removable media (Secure Digital, Compact Flash, etc.). The worm can render the mobile device unstable.[50]

In April 2005, Cambridge University security researchers published results of their actual implementation of passive attacks against the PIN-based pairing between commercial Bluetooth devices, confirming the attacks to be practicably fast and the Bluetooth symmetric key establishment method to be vulnerable. To rectify this vulnerability, they carried out an implementation which showed that stronger, asymmetric key establishment is feasible for certain classes of devices, such as mobile phones.[51]

In June 2005, Yaniv Shaked [52] and Avishai Wool [53] published a paper describing both passive and active methods for obtaining the PIN for a Bluetooth link. The passive attack allows a suitably equipped attacker to eavesdrop on communications and spoof, if the attacker was present at the time of initial pairing. The active method makes use of a specially constructed message that must be inserted at a specific point in the protocol, to make the master and slave repeat the pairing process. After that, the first method can be used to crack the PIN. This attack's major weakness is that it requires the user of the devices under attack to re-enter the PIN during the attack when the device prompts them to. Also, this active attack probably requires custom hardware, since most commercially available Bluetooth devices are not capable of the timing necessary.[54]

In August 2005, police in Cambridgeshire, England, issued warnings about thieves using Bluetooth enabled phones to track other devices left in cars. Police are advising users to ensure that any mobile networking connections are de-activated if laptops and other devices are left in this way.[55]

2006

In April 2006, researchers from Secure Network and F-Secure published a report that warns of the large number of devices left in a visible state, and issued statistics on the spread of various Bluetooth services and the ease of spread of an eventual Bluetooth worm.[56]

2007

In October 2007, at the Luxemburgish Hack.lu Security Conference, Kevin Finistere and Thierry Zoller demonstrated and released a remote root shell via Bluetooth on Mac OS X v10.3.9 and v10.4. They also demonstrated the first Bluetooth PIN and Linkkeys cracker, which is based on the research of Wool and Shaked.

Health concerns

Bluetooth uses the microwave radio frequency spectrum in the 2.402 GHz to 2.480 GHz range.[40] Maximum power output from a Bluetooth radio is 100 mW, 2.5 mW, and 1 mW for Class 1, Class 2, and Class 3 devices respectively, which puts Class 1 at roughly the same level as mobile phones, and the other two classes much lower.[57] Accordingly, Class 2 and Class 3 Bluetooth devices are considered less of a potential hazard than mobile phones, and Class 1 may be comparable to that of mobile phones : the maximum for a Class 1 is 100 mW for Bluetooth but 250 mW for UMTS W-CDMA, 1 W for GSM1800/1900 and 2 W for GSM850/900 for instance.

Bluetooth Innovation World Cup marketing initiative

The *Bluetooth* Innovation World Cup is an international competition encouraging the development of new innovations and ideas for applications leveraging the *Bluetooth* low energy wireless technology in sports, fitness and health care products.The *Bluetooth* Innovation World Cup is a marketing initiative of the Bluetooth Special Interest Group (SIG).

The aim of the competition is to stimulate new markets, creating new fields of applications and establishing Bluetooth low energy technology as the wireless data transfer standard for low energy applications is ordinary business in the competition of wireless standards. The initiative will go on for three years, having started 1 June 2009.[58]

Bluetooth Innovation World Cup 2009

The first international *Bluetooth* Innovation World Cup 2009 drew more than 250 international entries illustrating the abundance of opportunities for product development with the new *Bluetooth* low energy wireless technology.

The *Bluetooth* Innovation World Cup 2009 was sponsored by Nokia, Freescale Semiconductor, Texas Instruments, Nordic Semiconductor, STMicroelectronics and Brunel.

Bluetooth Innovator of the Year 2009

On February 8, 2010, the Bluetooth SIG has awarded Edward Sazonov, Physical Activity Innovations LLC, the title of *Bluetooth* Innovator of the Year for 2009. Sazonov received this recognition at the official award ceremony held in-line with the Wearable Technologies Show at ispo 2010, the world's largest trade show for sporting goods. The award includes a cash prize of €5,000 and a *Bluetooth* Qualification Program voucher (QDID) valued at up to US$ 10,000. Sazonov's winning idea, The Fit Companion, is a small, unobtrusive sensor that when clipped-on to a user's clothing or integrated in to a shoe, provides feedback about their physical activity. The data, transmitted via Bluetooth low energy technology, can help individuals to lose weight and achieve optimal physical activity. Intended for use in both training and daily activities like walking or performing chores, this simple, measuring device may offer a solution for reducing obesity.

Bluetooth Innovation World Cup 2010

The Bluetooth Special Interest Group (SIG) announced the start of the second *Bluetooth* Innovation World Cup (IWC) on 1 June 2010. The 2010 *Bluetooth* Innovation World Cup has a focus on applications for the sports & fitness, health care and home information and control markets. The competition will close for registrations on September 15, 2010.

See also

- Bluesniping
- BlueSoleil - Proprietary driver
- Continua Health Alliance
- DASH7
- Java APIs for Bluetooth
- Jellingspot Data Server
- Near Field Communication
- Tethering
- ZigBee - low power lightweight wireless protocol in the ISM band.

References

[1] "Bluetooth traveler" (http://www.hoovers.com/business-information/--pageid__13751--/global-hoov-index.xhtml). www.hoovers.com. . Retrieved 9 April 2010.

[2] Monson, Heidi (1999-12-14). "Bluetooth Technology and Implications" (http://www.sysopt.com/features/network/article.php/3532506). SysOpt.com. . Retrieved 2009-02-17.

[3] "About the Bluetooth SIG" (http://www.bluetooth.com/Bluetooth/SIG/). Bluetooth SIG. . Retrieved 2008-02-01.

[4] Kardach, Jim (2008-05-03). "How Bluetooth got its name" (http://www.eetimes.eu/scandinavia/206902019?cid=RSSfeed_eetimesEU_scandinavia). . Retrieved 2009-02-24.

[5] Newton, Harold. (2007). *Newton's telecom dictionary*. New York: Flatiron Publishing.

[6] "How Bluetooth Technology Works" (http://web.archive.org/web/20080117000828/http://bluetooth.com/Bluetooth/Technology/Works/). Bluetooth SIG. Archived from the original (http://www.bluetooth.com/Bluetooth/Technology/Works/) on 2008-01-17. . Retrieved 2008-02-01.

[7] "Class 1 Bluetooth Dongle Test" (http://www.amperordirect.com/pc/r-electronic-resource/z-reference-bluetooth-class1-myth.html). Amperordirect.com. . Retrieved 2010-09-04.

[8] "Profiles Overview" (http://www.bluetooth.com/English/Technology/Works/Pages/Profiles_Overview.aspx). Bluetooth.com. . Retrieved 2010-09-04.

[9] "Hypertag.com" (http://www.hypertag.com/company1/what-is-proximity-or-bluetooth-marketing/). Hypertag.com. . Retrieved 2010-09-04.

[10] "Wii Controller" (http://web.archive.org/web/20080220080315/http://bluetooth.com/Bluetooth/Products/Products/Product_Details.htm?ProductID=2951). Bluetooth SIG. Archived from the original (http://bluetooth.com/Bluetooth/Products/Products/Product_Details.htm?ProductID=2951) on February 20, 2008. . Retrieved 2008-02-01.

[11] "Telemedicine.jp" (http://www.telemedicine.jp/). Telemedicine.jp. . Retrieved 2010-09-04.

[12] "Real Time Location Systems" (http://www.clarinox.com/docs/whitepapers/RealTime_main.pdf). clarinox. . Retrieved 2010-8-4.

[13] http://www.bluetooth.com/English/Products/Pages/Watch.aspx

[14] "Specification Documents" (http://www.bluetooth.com/Specification Documents/AssignedNumbersServiceDiscovery.pdf). 30-06-2010. .

[15] Apple (2002-07-17). "Apple Introduces "Jaguar," the Next Major Release of Mac OS X" (http://www.apple.com/pr/library/2002/jul/17jaguar.html). Press release. . Retrieved 2008-02-04.

[16] "Bluetooth Wireless Technology FAQ - 2010" (http://download.microsoft.com/download/9/c/5/9c5b2167-8017-4bae-9fde-d599bac8184a/Bth_FAQ.docx). . Retrieved 2010-09-04.

[17] "Network Protection Technologie" (http://www.microsoft.com/technet/prodtechnol/winxppro/maintain/sp2netwk.mspx). *Changes to Functionality in Microsoft Windows XP Service Pack 2*. Microsoft Technet. . Retrieved 2008-02-01.

[18] "Official Linux Bluetooth protocol stack" (http://www.bluez.org/). BlueZ. . Retrieved 2010-09-04.

[19] OMTP.org (http://www.omtp.org/Publications/Display.aspx?Id=8f152a02-4120-4933-a1e5-74c7ad472bc8)

[20] "The Bluetooth Blues" (http://web.archive.org/web/20071222231740/http://www.information-age.com/article/2001/may/the_bluetooth_blues). Information Age. 2001-05-24. Archived from the original (http://www.information-age.com/article/2001/may/the_bluetooth_blues) on 2007-12-22. . Retrieved 2008-02-01.

[21] "IEEE Std 802.15.1-2002 – IEEE Standard for Information technology – Telecommunications and information exchange between systems – Local and metropolitan area networks – Specific requirements Part 15.1: Wireless Medium Access Control (MAC) and Physical Layer (PHY) Specifications for Wireless Personal Area Networks (WPANs)" (http://ieeexplore.ieee.org/servlet/opac?punumber=7932). Ieeexplore.ieee.org. doi:10.1109/IEEESTD.2002.93621. . Retrieved 2010-09-04.

[22] Guy Kewney (2004-11-16). "High speed Bluetooth comes a step closer: enhanced data rate approved" (http://www.newswireless.net/index.cfm/article/629). Newswireless.net. . Retrieved 2008-02-04.

[23] "IEEE Std 802.15.1-2005 – IEEE Standard for Information technology – Telecommunications and information exchange between systems – Local and metropolitan area networks – Specific requirements Part 15.1: Wireless Medium Access Control (MAC) and Physical Layer (PHY) Specifications for Wireless Personal Area Networks (WPANs)" (http://ieeexplore.ieee.org/servlet/opac?punumber=9980). Ieeexplore.ieee.org. doi:10.1109/IEEESTD.2005.96290. . Retrieved 2010-09-04.
[24] "Specification Documents" (http://www.bluetooth.com/Bluetooth/Technology/Building/Specifications/). Bluetooth SIG. . Retrieved 2008-02-04.
[25] "HTC TyTN Specification" (http://www.europe.htc.com/z/pdf/products/1766_TyTN_LFLT_OUT.PDF) (PDF). HTC. . Retrieved 2008-02-04.
[26] (PDF) *Simple Pairing Whitepaper* (http://web.archive.org/web/20061018032605/http://www.bluetooth.com/NR/rdonlyres/0A0B3F36-D15F-4470-85A6-F2CCFA26F70F/0/SimplePairing_WP_V10r00.pdf). Version V10r00. Bluetooth SIG. 2006-08-03. Archived from the original (http://bluetooth.com/NR/rdonlyres/0A0B3F36-D15F-4470-85A6-F2CCFA26F70F/0/SimplePairing_WP_V10r00.pdf) on October 18, 2006. . Retrieved 2007-02-01.
[27] https://www.bluetooth.org/apps/content/
[28] David Meyer (2009-04-22). "Bluetooth 3.0 released without ultrawideband" (http://news.zdnet.co.uk/communications/0,1000000085,39643174,00.htm). zdnet.co.uk. . Retrieved 2009-04-22.
[29] http://www.bluetooth.com/SiteCollectionDocuments/Core_V30HS.zip
[30] Nokia (2007-06-12). "Wibree forum merges with Bluetooth SIG" (http://www.wibree.com/press/Wibree_pressrelease_final_1206.pdf) (PDF). Press release. . Retrieved 2008-02-04.
[31] "Bluetooth.com" (http://www.bluetooth.com/Bluetooth/Press/SIG/SIG_INTRODUCES_BLUETOOTH_LOW_ENERGY_WIRELESS_TECHNOLOGY_THE_NEXT_GENERATION_OF_BLUETOOTH_WIRELESS_TE.htm). Bluetooth.com. . Retrieved 2010-09-04.
[32] "Wimedia.org" (http://www.wimedia.org/). Wimedia.org. 2010-01-04. . Retrieved 2010-09-04.
[33] "Wimedia.org" (http://www.wimedia.org/imwp/download.asp?ContentID=15508). Wimedia.org. . Retrieved 2010-09-04.
[34] "Wimedia.org" (http://www.wimedia.org/imwp/download.asp?ContentID=15506). . Retrieved 2010-09-04.
[35] Bluetooth.com (http://www.bluetooth.com/Bluetooth/Technology/Technology_Transfer/)
[36] "USB.org" (http://www.usb.org/press/WiMedia_Tech_Transfer/). USB.org. 2009-03-16. . Retrieved 2010-09-04.
[37] "Incisor.tv" (http://www.incisor.tv/2009/03/what-to-make-of-bluetooth-sig-wimedia.html). Incisor.tv. 2009-03-16. . Retrieved 2010-09-04.
[38] Bluetooth group drops ultrawideband, eyes 60 GHz (http://www.eetimes.com/showArticle.jhtml;jsessionid=J5E0PN3NQ5BNLQE1GHPSKH4ATMY32JVN?articleID=221100170), Report: Ultrawideband dies by 2013 (http://www.eetimes.com/showArticle.jhtml;jsessionid=J5E0PN3NQ5BNLQE1GHPSKH4ATMY32JVN?articleID=217201265), Incisor Magazine November 2009 (http://www.incisor.tv/download.php?file=140november2009.pdf)
[39] Stallings, William. (2005). *Wireless communications & networks.'=' Upper Saddle River, NJ: Pearson Prentice Hall.*
[40] D. Chomienne, M. Eftimakis (2010-10-20). "Bluetooth Tutorial" (http://www.newlogic.com/products/Bluetooth-Tutorial-2001.pdf) (PDF). . Retrieved 2009-12-11.
[41] Juha T. Vainio (2000-05-25). "Bluetooth Security" (http://www.iki.fi/jiitv/bluesec.pdf). Helsinki University of Technology. . Retrieved 2009-01-01.
[42] Andreas Becker (2007-08-16) (PDF). *Bluetooth Security & Hacks* (http://gsyc.es/~anto/ubicuos2/bluetooth_security_and_hacks.pdf). Ruhr-Universität Bochum. . Retrieved 2007-10-10.
[43] Scarfone, K., and Padgette, J. (September 2008) (PDF). *Guide to Bluetooth Security* (http://csrc.nist.gov/publications/nistpubs/800-121/SP800-121.pdf). National Institute of Standards and Technology. . Retrieved 2008-10-03.
[44] "What is bluejacking?" (http://www.bluejackq.com/what-is-bluejacking.shtml). Helsinki University of Technology. . Retrieved 2008-05-01.
[45] "Security Weaknesses in Bluetooth" (http://citeseerx.ist.psu.edu/viewdoc/summary?doi=10.1.1.23.7357). RSA Security Conf. – Cryptographer's Track. . Retrieved 2009-03-01.
[46] "Bluetooth" (http://web.archive.org/web/20070126012417/http://www.thebunker.net/resources/bluetooth). The Bunker. Archived from the original (http://www.thebunker.net/resources/bluetooth) on January 26, 2007. . Retrieved 2007-02-01.
[47] "BlueBug" (http://trifinite.org/trifinite_stuff_bluebug.html). Trifinite.org. . Retrieved 2007-02-01.
[48] John Oates (2004-06-15). "Virus attacks mobiles via Bluetooth" (http://www.theregister.co.uk/2004/06/15/symbian_virus/). The Register. . Retrieved 2007-02-01.
[49] "Long Distance Snarf" (http://trifinite.org/trifinite_stuff_lds.html). Trifinite.org. . Retrieved 2007-02-01.
[50] "F-Secure Malware Information Pages: Lasco.A" (http://www.f-secure.com/v-descs/lasco_a.shtml). F-Secure.com. . Retrieved 2008-05-05.
[51] Ford-Long Wong, Frank Stajano, Jolyon Clulow (2005-04) (PDF). *Repairing the Bluetooth pairing protocol* (http://web.archive.org/web/20070616082657/http://www.cl.cam.ac.uk/~fw242/publications/2005-WongStaClu-bluetooth.pdf). University of Cambridge Computer Laboratory. Archived from the original (http://www.cl.cam.ac.uk/~fw242/publications/2005-WongStaClu-bluetooth.pdf) on 2007-06-16. . Retrieved 2007-02-01.
[52] http://www.eng.tau.ac.il/~shakedy
[53] http://www.eng.tau.ac.il/~yash/

[54] Yaniv Shaked, Avishai Wool (2005-05-02). *Cracking the Bluetooth PIN* (http://www.eng.tau.ac.il/~yash/shaked-wool-mobisys05/). School of Electrical Engineering Systems, Tel Aviv University. . Retrieved 2007-02-01.
[55] "Phone pirates in seek and steal mission" (http://web.archive.org/web/20070717035938/http://www.cambridge-news.co.uk/news/region_wide/2005/08/17/06967453-8002-45f8-b520-66b9bed6f29f.lpf). Cambridge Evening News. Archived from the original (http://www.cambridge-news.co.uk/news/region_wide/2005/08/17/06967453-8002-45f8-b520-66b9bed6f29f.lpf) on 2007-07-17. . Retrieved 2008-02-04.
[56] (PDF) *Going Around with Bluetooth in Full Safety* (http://www.securenetwork.it/bluebag_brochure.pdf). F-Secure. 2006-05. . Retrieved 2008-02-04.
[57] M. Hietanen, T. Alanko (2005-10). "Occupational Exposure Related to Radiofrequency Fields from Wireless Communication Systems" (http://web.archive.org/web/20061006124651/http://www.ursi.org/Proceedings/ProcGA05/pdf/K03.7(01682).pdf) (PDF). *XXVIIIth General Assembly of URSI - Proceedings*. Union Radio-Scientifique Internationale (http://www.ursi.org/). Archived from the original (http://www.ursi.org/Proceedings/ProcGA05/pdf/K03.7(01682).pdf) on October 6, 2006. . Retrieved 2007-04-19.
[58] "Bluetooth Innovation World Cup" (http://www.bluetooth.com/Bluetooth/Press/Bluetooth_World_Innovation_Cup.htm). Bluetooth.com. . Retrieved 2010-09-04.

External links

- Official website (http://http://www.bluetooth.org/)
- Use cases for mobile phones (http://www.bluetooth.com/English/Experience/Pages/On_Your_Phone.aspx)

Sony Ericsson

Sony Ericsson

Type	Joint venture
Industry	Telecommunications
Founded	October 1, 2001[1]
Headquarters	Hammersmith, London, United Kingdom
Area served	Worldwide
Key people	Sir Howard Stringer (Chairman) Bert Nordberg (President) Rikko Sakaguchi (EVP) Kristian Tear (EVP) William A Glaser Jr (CFO)[2]
Products	Mobile phones Mobile music devices Wireless systems Wireless voice devices Hi-Tech accessories Wireless data devices
Revenue	▼ €6.788 billion (2009)[3]
Profit	▼ -€836 million (2009)
Employees	8,450 (as of April 2010)[4]
Parent	Sony Corporation (50%) Ericsson AB (50%)
Website	SonyEricsson.com [5]

Sony Ericsson is a joint venture established on October 1, 2001[1] by the Japanese consumer electronics company Sony Corporation and the Swedish telecommunications company Ericsson to manufacture mobile phones. The stated reason for this venture is to combine Sony's consumer electronics expertise with Ericsson's technological knowledge in the communications sector. Both companies have stopped making their own mobile phones.

The company's global management is based in Hammersmith in London, United Kingdom, and it has research & development teams in Lund, Sweden; Tokyo, Japan; Beijing, China and Redwood Shores, United States. By 2009, it was the fourth-largest mobile phone manufacturer in the world after Nokia, Samsung and LG.[6] The sales of products largely increased due to the launch of the adaptation of Sony's popular Walkman and Cyber-shot series.

Recent performance

While Sony Ericsson has been enjoying strong growth recently, its South Korean rival LG Electronics overtook it in Q1 2008 due to the company's profits falling significantly by 43% to €133 million (approx. US$ 179.6697 million[7]), sales falling by 8% and market share dropping from 9.4% to 7.9%, despite favourable conditions that the handset market was expected to grow by 10% in 2008. Sony Ericsson announced another profit warning in June 2008[8] and saw net profit crash by 97% in Q2 2008, announcing that it would cut 2,000 jobs, leading to wide fear that Sony Ericsson is on the verge of decline along with its struggling rival, Motorola.[9] In Q3 the profits were much on the same level, however November and December saw increased profits along with new models being released such as the C905 being one of the top sellers across the United Kingdom.

Sony Ericsson has, as of July 18, 2008, approximately 9,400 employees and 2,500 contractors worldwide. Bert Nordberg is the president of the company since October, 2009. Sir Howard Stringer, CEO and President, Sony Corporation, is chairman of the board.

History

Troubles in Ericsson's mobile phone business

In the United States, Ericsson partnered with General Electric in the early nineties, primarily to establish a US presence and brand recognition.

Ericsson had decided to obtain chips for its phones from a single source—a Philips facility in New Mexico. In March 2000, a fire at the Philips factory contaminated the sterile facility. Philips assured Ericsson and Nokia (their other major customer) that production would be delayed for no more than a week. When it became clear that production would actually be compromised for months, Ericsson was faced with a serious shortage. Nokia had already begun to obtain parts from alternative sources, but Ericsson's position was much worse as production of current models and the launch of new ones was held up.[10]

Ericsson, which had been in the cellular phone market for decades, and was the world's no. 3 cellular telephone handset maker, was struggling with huge losses. This was mainly due to this fire and its inability to produce cheaper phones like Nokia. To curtail the losses, it considered outsourcing production to Asian companies that could produce the handsets for lower costs.

Speculation began about a possible sale by Ericsson of its mobile phone division, but the company's president said they had no plans to do so. "Mobile phones are really a core business for Ericsson. We wouldn't be as successful (in networks) if we didn't have phones", he said.

Background of the joint venture

Sony was a marginal player in the worldwide cell phone market with a share of less than 1 percent in 2000.

By August 2001, the two companies had finalized the terms of the merger announced in April. The company was to have an initial workforce of 3,500 employees.

Ericsson's market share actually fell and in August 2002, Ericsson said it would stop making mobile phones and end its partnership with Sony if the business continued to disappoint However, in January 2003, both companies said they would inject more money into the joint venture in a bid to stem the losses.

Sony Ericsson's strategy was to release new models capable of digital photography as well as other multimedia capabilities such as downloading and viewing video clips and personal information management capabilities. To this end, it released several new models which had built-in digital camera and color screen which were novelties at that time. The joint venture, however, continued to make bigger losses in spite of booming sales. The target date for making a profit from its first year to 2002 was postponed to 2003 to second half of 2003. It failed in its mission of becoming the top seller of multimedia handsets and was in fifth-place and struggling in 2005.

Turnaround

Beginning of the turnaround

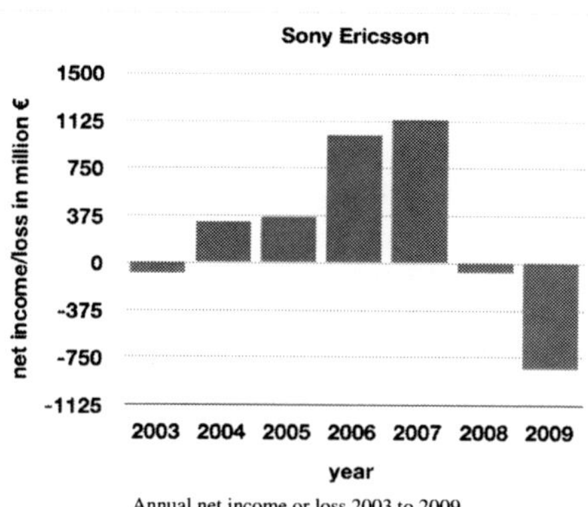

Annual net income or loss 2003 to 2009

In June 2002, Sony Ericsson announced it would stop making Code division multiple access (CDMA) cellphones for the US market and would focus on GSM as the dominant technology. It also cut jobs in research and development in the USA and in Germany. In October 2003, it posted its first quarterly profit but warned that falling prices on phones and competition would make it difficult to stay in the black. Sony Ericsson's recovery is credited to the success of the T610 model. Although Sony Ericsson no longer makes CDMA cellphones for the global market, they still manufacture CDMA cellphones for Japanese market, mainly for au by KDDI.

Following the success of its P800 phone, Sony Ericsson introduced the P900 at simultaneous events in Las Vegas and Beijing in October 2003. It was pegged as smaller, faster, simpler and more flexible than its predecessor.[11]

In March 2004, Ericsson said it would try to block its rival Nokia from gaining control of Symbian, an industry consortium that made operating software for smart phones.

In 2004, Sony Ericsson's market share increased from 5.6 percent in the first quarter to 7 percent in the second quarter . In July 2004, Sony Ericsson unveiled the P910 communicator with its integrated thumbboard, broad e-mail support, quadruple memory and improved screen.

In February 2005, Sony Ericsson president Miles Flint announced at the 3GSM World Congress that Sony Ericsson would unveil a mobile phone/digital music player in the next month. It would be called the Walkman phone and would play music file formats such as MP3 and AAC.

Success with Walkman and Cyber-shot phones

On March 1, 2005, Sony Ericsson introduced the K750i with a 2 megapixel camera, as well as its platform mate, the W800i, the first of the Walkman phones capable of 30 hours of music playback, and two low-end phones.

In 2007 their first 5-Megapixel camera phone, the Sony Ericsson K850i, was announced followed in 2008 by the Sony Ericsson C905, the world's first 8-Megapixel phone. At Mobile World Congress 2009, Sony Ericsson unveiled the first 12-Megapixel phone, named Satio, on 28 May 2009.

Beyond

On May 1, 2005, Sony Ericsson agreed to become the global title sponsor for the WTA Tour in a deal worth 88 million US dollars over 6 years. The women's pro tennis circuit was renamed the Sony Ericsson WTA Tour. Just over a month later on June 7, it announced sponsorship of West Indian batsmen Chris Gayle and Ramnaresh Sarwan.

In October 2005, Sony Ericsson presented the first mobile phone based on UIQ 3, the P990.

On January 2, 2007, Sony Ericsson announced in Stockholm that it will be having some of its mobile phones produced in India. It announced that its two outsourcing partners, FLextronics and Foxconn will be producing 10 million cellphones per year by 2009. CEO Miles Flint announced at a press conference held with India's communications minister Dayanidhi Maran in Chennai that India was one of the fastest growing markets in the world and a priority market for Sony Ericsson with 105 million users of GSM mobile telephones.

On February 2, 2007, Sony Ericsson acquired UIQ Technology, a Swedish software company from Symbian Ltd.. UIQ will remain an independent company, Miles Flint announced.[12]

On October 15, 2007, Sony Ericsson announced on Symbian Smartphone Show that they will be selling half of its UIQ share to Motorola thus making UIQ technology owned by two large mobile phone companies.

Sony Ericsson Cost-Cutting Program And Job Losses

In June 2008, Sony Ericsson had about 12,000 employees, it then launched a cost-cutting program and by the end of 2009 it had slashed its global workforce by around 5,000 people. It planned to cut another 1,500 jobs in 2010. It has also closed R&D (research and development) centres globally, such as, Chadwick House [13], Birchwood (Warrington) in the UK; Miami, Seattle, San Diego and RTP (Raleigh, NC) in the USA; The Chennai Unit (New Delhi) in India; Hässleholm and Kista in Sweden and operations in the Netherlands. The UIQ centres in London and Budapest were also closed, UIQ was a joint venture with Motorola which began life in the 1990s.[14] [15] [16] [17] [18] [19] [20] [21] [22] [23] [24]

Types of phones

Main areas of interest

Sony Ericsson currently concentrates on the categories of: music, camera, business (web and email), design, all-rounder, eco-friendly, and budget focused phones. Its six largest categories are:

The Sony Ericsson K750i.

- The Walkman-branded W series music phones, launched in 2005.
  The Sony Ericsson W-series music phones are notable for being the first music-centric series mobile phones, prompting a whole new market for portable music that was developing at the time. The main feature that can be seen in all of these Walkman phones is they all have a 'W' button which once pressed opens the media center. Sony Ericsson's Walkman phones have previously been commercially endorsed by pop stars Christina Aguilera and Jason Kay across Europe. The latest model, Yendo, is the first Walkman branded phone with full touchscreen. Walkman branded phones are also produced for the Japanese market, where one Walkman branded phone launched in 2009, Premier³ (Premier Cube), is able to rip music from a CD player directly to the phone via a connector.
- The Cyber-shot-branded line of phones, launched in 2006 in newer models of the K series phones.
  This range of phones are focused on the quality of the camera included with the phone. Cyber-shot phones always include a flash, some with a xenon flash, and also include auto-focus cameras. Sony Ericsson kicked off its global marketing campaign for Cyber-shot phone with the launch of 'Never Miss a Shot'. The campaign featured top female tennis players Ana Ivanović and Daniela Hantuchová. On 10 February 2008, the series has been expanded with the announcement of C702, C902 and C905 phones. The latest model, S003, which launched in 2010 for the Japanese market only, is the second of Cyber-shot branded phone series which features a 12 Megapixel camera after Satio. It uses 'Exmor' CMOS sensor, Dual-LED 'PLASMA' Flash, ISO 3200, and is waterproof.
- The BRAVIA-branded line of phones, launched 2007 in the Japanese market only.
  Until now, four BRAVIA branded phones have been produced. Sony Ericsson (FOMA SO903iTV, FOMA SO906i, U1, and S004[25]) uses the BRAVIA brand. BRAVIA branded phone are able to show 1seg terrestrial television.

- The UIQ smartphone range of mobiles, introduced with the P series in 2003 with the introduction of P800. They are notable for their touchscreens, QWERTY keypads (on most models), and use of the UIQ interface platform for Symbian OS. This range has since expanded into the M series and G series phones.
- The XPERIA range of mobile phones, heralded by the Sony Ericsson XPERIA X1 on February 2008 at the Mobile World Congress (formerly 3GSM) held in Barcelona Spain, was the first trademark promoted by the Sony Ericsson as its own and is designated to provide technological convergence among its target user base. The first model, X1, carried the Windows Mobile operating system with a Sony Ericsson's panel interface. The Xperia X10 model features the Android operating system. Additionally, Yahoo! News reported that Sony will align with Google to run Android on its upcoming gaming smartphone.[26]
- The GreenHeart range of mobile phones, first introduced in 2009, heralded by the Sony Ericsson J105i Naite and C901 GreenHeart.
  It is focused on an environmentally friendly theme, but still featured with recent mobile technology and multimedia capability. It mainly uses eco-friendly materials and features eco-apps.

Phone series description

Series	Branding	Description	Origin
C	Cyber-shot	Camera-focused phones.	Cyber-shot
D	T-Mobile	T-Mobile network exclusive phones.	Deutsche Telekom
F	Vodafone (partial)	Vodafone network exclusive phones; Gaming focused phones	Vodafone / Fun
G	Web	Web browsing-oriented phones.	Generation Web
J	Junior/Low end	Low-end series	Junior
K	Cyber-shot (partial) and QuickShare (partial)	All-around phones	Kamera, *Swedish for camera*
M	QWERTY, fashion-focused business phones	Business-focused smartphones.	Messaging
P	PDA UIQ (Symbian) phones	Powerhouse smartphones.	PDA
R	Radio	Low end phones that are made for AM/FM radio	Radio
S	Style/Sagem	Fashion low end phones, mostly part produced by Sagem	Swivel, Slider, Snapshot, Sagem
T	Talker/Older Ericsson	All-around phones/Older Ericsson handsets	Tala *(Swedish for "talk")*
TM	T-Mobile	T-Mobile USA network exclusive phones	T-Mobile
U	Entertainment Unlimited	Internal model number for phones launched under the "Entertainment Unlimited" brand	Unlimited
V	Vodafone	Vodafone network exclusive phones.	Vodafone
W	Walkman	Music-focused phones.	Walkman
X	Xperia	Convergence and powerhouse devices.	Xperia
Z	Ze Bobber (meaning With Flip)	Design-oriented phones/clamshells	Ze Bobber

Naming convention

Current system

After the 2008 Mobile World Congress, Sony Ericsson announced their new naming system comprising four characters, each character denoting the "Series", the "Range/Class", the "Version" and the "Form Factor" respectively.

Series	Range/Class	Version	Form Factor
(see above for series letters)	1-4: Low-end 5-7: Mid-range 8-9: High-end	(in numerical order of succession)	0-2: Candybar 3-5: Slider 6-8: Clamshell 9: Others

Some older Sony Ericsson handsets and some older that are still on the market (i.e. Sony Ericsson W890i) use suffixes for the market it is sold in, or stripped down features. The suffixes used were:

- A000**a** which meant for American/US market only.
- A000**i** was for the Asia-pacific region
- A000**im** was for i-mode mobile phones by Sony Ericsson
- A000**c** was for the Chinese market only.

All new devices created by Sony Ericsson will not use suffixes but just the model (i.e. instead of **A000a** there will be **A000** in every market, to avoid confusion, they will also have names replaced instead of a model to avoid even more confusion (i.e. instead of Sony Ericsson U10i they named it Sony Ericsson Aino)

Previous systems

Sony Ericsson has used three methods in the past of naming their mobile products:

- The most common format uses a total of five (or six) characters, e.g. K750i.
  This format begins with a capital letter to denote the series of the phone (**K**750i). This is then followed by three numbers (K**750**i). The first number indicates the sub-series of the phone, the second indicates the amount of progression from the previous release, i.e. K**7**00i to K7**5**0i, and the third number is always either a '0','5' or '8'. The number '8' is used either to show a variation of the phone destined for a different market without a feature, e.g. the W88**8** is a W880i without 3G, or it is used to separate phones which have identical specifications but the designs are different, e.g. K61**0**i and the K61**8**i or k800i and the k810i. The number '5' is used for newer models, where the first two numbers and the zero have already been used in a previous model, for example in the case of the W70**0** and the W70**5**, which allows for more naming options. Finally, the lowercase letter at the end of the model name describes the market for which a product is intended; these are: **a** for the Americas, **c** for China, and **i** stands for an international version; there is also an 'im' suffix used for branding i-mode phones. Often the last letter is left out to describe the phone generically with no region specific branding.
- A newer format uses a total of three characters, e.g. P1i. It is believed that this format is intended for naming flagship models of each phone series due to the limited numbering combinations.
  It begins with a capital letter to denote the series of the phone (**P**1i). The number is used to indicate the amount of progression from the previous release (P**1**i) and the final lowercase letter, as explained above, describes the market for which a product is intended. Again, often the last letter is left out to describe the phone generically with no region specific branding.
- The oldest naming format uses a total of four characters, e.g. T68i. This format continued from the naming scheme of the Ericsson mobile business and was only ever used once.
  This format begins with a capital letter to denote the series of the phone (**T**68i). The first number indicates the

sub-series of the phone (T**6**8i) and the second letter indicates the amount of progression from the previous release. The last lowercase letter indicates that it is an update of the previous model, i.e. T68 to T68**i**.

Another peculiar naming format was the one used in naming the Z1010; this format has not been used since the Z1010.

Furthermore, Sony Ericsson always give their phones codenames when developing. Mainly to keep the information secret and to prevent leaks. All codenames are female names, and some have been taken from the female players of the Sony Ericsson-sponsored tennis tournament, WTA.

Financial information

Sony Ericsson posted its first profit in the second half of 2003. Since then, the sales figures from phones have been:

- 2004: 42 million units[27]
- 2005: 50 million units[28]
- 2006: 74.8 million units[29]
- 2007: 103.4 million units[30]
- 2008: 96.6 million units[31]
- 2009: 57.1 million units[32]

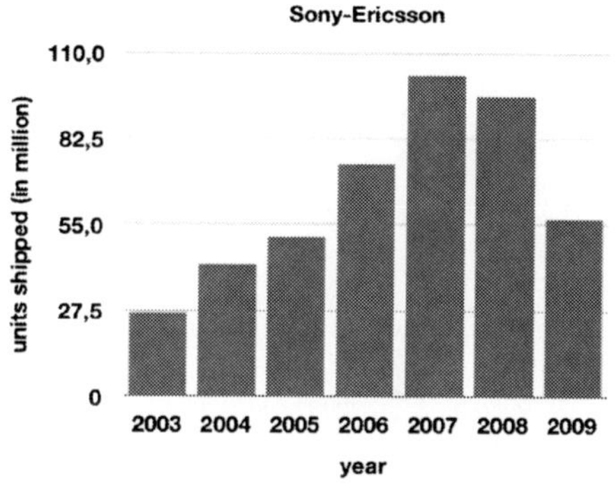

Annual shipments of units 2003 to 2009

According to the Swedish Magazine *M3*s issue 7/2006 Sony Ericsson is the best-selling phone brand in the Nordic countries, followed by Nokia.

In the third quarter of 2009, Sony Ericsson became the world's fourth largest mobile phone manufacturer with 4.9% of market share after Nokia(37.8%), Samsung(21%) and LG(11%).

Compatibility

During E3 2007 Media and Business Summit, Phil Harrison, Sony CEO showcased a Sony Ericsson phone using the PlayStation's XMB. A select group of phones are also said to integrate into PlayStation Home *(final product)*

During the announcement of Sony Ericsson K850, W960 and W910 some review sites have shown that those mentioned phones and future mid-range or better phones will have Media to replace the standard File Manager which will be moved to the Organiser of the phone. The Media manager possesses a UI that resembles the XMB interface found on Sony TV, PS3, and PSP products. The mobile developer site confirmed from their spec sheets and white papers that the XMB media manager is standard to the phones running Java Platform 8 also known as A200 Platform and Symbian devices like Sony Ericsson Satio and Sony Ericsson Vivaz.

Operations

In 2009 Sony Ericsson announced that it was moving its North American headquarters from Research Triangle Park, North Carolina to Atlanta. The headquarters move was part of a plan to reduce its workforce, then 10,000 employees, by 20%. As of that year Sony Ericsson had 425 employees in Research Triangle Park; the staff had been reduced by hundreds due to layoffs.[33] Stacy Doster, a spokesperson of Sony Ericsson, said that the proximity to Hartsfield-Jackson Atlanta International Airport's flights to Latin America and the operations of AT&T Mobility influenced the decision to move the USA headquarters. Sony Ericsson will close the Research Triangle site.[33] [34]

Environmental record

Sony Ericsson ranks 2nd behind Nokia out of 18 leading electronics makers in Greenpeace's Guide to Greener Electronics that assesses their policies on toxic chemicals, recycling and climate change. It is the first company to score full marks on all chemical criteria in the ranking.[35] Sony Ericsson is ahead of many of its competitors in eliminating chemical substances in its products. All Sony Ericsson products are free of toxic vinyl plastic (PVC) and brominated flame retardants (BFRs) except for the ongoing phase-out of a few components. It has removed antimony, beryllium and phthalate from models launched since 2008.[36] Greenpeace criticises Sony Ericssons limited take-back and recycling programme, as well as its limited use of recycled plastic in its products.[37] . However, in June 2009 launched its first GreenHeart series device, the C901, which indirectly emits a 15% less of CO_2 during its fabrication and usage, compared to other SE phones. It is also packed in a small box without paper manual, includes an eco-charger, and its cover is made of recycled plastic.

See also

- Sony
- Ericsson
- LG
- Nokia
- Samsung
- Symbian
- MyPhoneExplorer - full featured Sony Ericsson manager software for MS Windows, (able to backup, synch and editing almost all data on the phone via Bluetooth, infrared or USB)
- Media Go - music, photo, video, and game management software made for Sony Ericsson phones
- Disc2Phone - music management software made for Sony Ericsson phones
- SonicStage - music management software made for Japan market phones.
- PlayNow - global content distribution portal
- List of Sony Ericsson products

References

[1] "Ericsson - press release" (http://www.cisionwire.com/ericsson/sony-ericsson-mobile-communications-established-today). Cision Wire. . Retrieved 2001-10-01.

[2] http://www.sonyericsson.com/cws/corporate/company/aboutus/executivebiographies

[3] "Sony Ericsson Reports Fourth Quarter and Full Year 2009 Results" (http://www.sonyericsson.com/cws/corporate/press/pressreleases/pressreleasedetails/q409pressreleasefinal-20100122)

[4] IDG.se - 3150 have been fired (http://translate.google.com/translate?js=y&prev=_t&hl=sv&ie=UTF-8&layout=1&eotf=1&u=http://www.idg.se/2.1085/1.311091/3-150-har-fatt-sparken&sl=sv&tl=en)

[5] http://www.sonyericsson.com

[6] http://www.telecomskorea.com/market-8211.html

[7] €133 million to $ - Google Search (http://www.google.com/search?hl=en&safe=off&q=€133+million+to+$&btnG=Search)

[8] Sony Ericsson issues second profit warning of the year, hopes to break even in Q2 - Engadget (http://www.engadget.com/2008/06/30/sony-ericsson-issues-second-profit-warning-of-the-year-expects/)

[9] Sony Ericsson Profits Crash 48% - TrustedReviews (http://www.trustedreviews.com/mobile-phones/news/2008/04/24/Sony-Ericsson-Profits-Crash-48-/p1)

[10] *"When the chain breaks"* (June 17, 2006). The Economist: A survey of logistics, p. 18.

[11] Pilato, Fabrizio (October 20, 2003). "Sony Ericsson unveils the P900 Symbian OS Powered Smartphone" (http://www.mobilemag.com/2003/10/20/sony-ericsson-unveils-the-p900-symbian-os-powered-smartphone/). .

[12] Sony Ericsson seeks shareholders for UIQ software company - MarketWatch (http://www.marketwatch.com/news/story/sony-ericsson-seeks-shareholders-uiq/story.aspx?guid={4BE5DD6B-9BEE-4085-B3A8-3EAF9878B747})

[13] http://www.mepc.com/birchwoodpark/Home.aspx

[14] "Sony Ericsson to close down unit in Manchester" (http://www.cn-c114.net/578/a349096.html). cn-c114.net. .

[15] "UK redundancies reach 25,000 in just a week" (http://business.timesonline.co.uk/tol/business/economics/article5201539.ece). Times Online. .

[16] "China wins, Symbian loses in Sony Ericsson reorg" (http://www.theregister.co.uk/2008/09/30/sony_ericsson_reorg). The Register. .

[17] "Ericsson to cut a further 5,000 jobs" (http://www.telegraph.co.uk/finance/newsbysector/mediatechnologyandtelecoms/4305325/Ericsson-to-cut-a-further-5000-jobs.html). Telegraph. .

[18] "Sony Ericsson to Cut 2,000 More Jobs After Third Loss (Update3)" (http://www.bloomberg.com/apps/news?pid=newsarchive&sid=aB0MUA6dgOk4). Bloomberg. .

[19] "Sony Ericsson to lay off 2,000 more workers" (http://www.zdnet.co.uk/news/networking/2009/04/17/sony-ericsson-to-lay-off-2000-more-workers-39641353). ZDNet. .

[20] "With Profit Down 82% for Quarter, Ericsson Plans More Job Cuts" (http://www.nytimes.com/2010/01/26/technology/companies/26ericsson.html?_r=1). NY Times. .

[21] "Sony Ericsson to close Kista development centre" (http://www.thelocal.se/23350/20091119). The Local. .

[22] "Sony Ericsson to shut down Chennai unit" (http://www.thehindubusinessline.com/2009/11/20/stories/2009112052150100.htm). The Hindu Business Line. .

[23] "Four Facilities Closing and 2000 job Losses with Sony Ericsson" (http://www.phonesreview.co.uk/2009/11/18/four-facilities-closing-and-2000-job-losses-with-sony-ericsson). Phones Review. .

[24] "Sony Ericsson to close RTP site" (http://www.wral.com/business/story/6446521). WRAL. .

[25] NTT docomo SO906i (http://www.nttdocomo.co.jp/english/product/foma/906i/so906i/index.html)

[26] Mokey, Nick. "Sony Adopts Android 3.0 for Upcoming PlayStation Phone" (http://news.yahoo.com/s/digitaltrends/20100812/tc_digitaltrends/sonyadoptsandroid30forupcomingplaystationphone_1), Yahoo! News, 12 August 2010. Retrieved on 08-12-10.

[27] Gartner Says Top Six Vendors Drive Worldwide Mobile Phone Sales to 21 Percent Growth in 2005 (http://www.gartner.com/it/page.jsp?id=492248), gartner.com, 28 February 2006.

[28] Sony Ericsson reports record shipments, sales and profits (http://www.sonyericsson.com/spg.jsp?cc=global&lc=en&ver=4001&template=pc3_1_1&zone=pc&lm=pc3&prid=4532), sonyericsson.com, 18 January 2006.

[29] Record quarter caps a record year for Sony Ericsson (http://www.sonyericsson.com/spg.jsp?cc=global&lc=en&ver=4001&template=pc3_1_1&zone=pc&lm=pc3&prid=7012), sonyericsson.com, 17 January 2007.

[30] Sony Ericsson sells over 100 million handsets in 2007 (http://www.sonyericsson.com/cws/corporate/press/pressreleases/pressreleasedetails/q42007financialpressreleases-20080116), sonyericsson.com, 16 January 2008.

[31] Sony Ericsson reports results for fourth quarter and full year 2008 (http://www.sonyericsson.com/cws/corporate/press/pressreleases/pressreleasedetails/pressreleaseq408-20090116), sonyericsson.com, 16 January 2009.

[32] Sony Ericsson reports results for fourth quarter and full year 2009 (http://www.sonyericsson.com/cws/corporate/press/pressreleases/pressreleasedetails/q409pressreleasefinal-20100122), sonyericsson.com, 22 January 2010.

[33] Dalesio, Emery P. " Sony Ericsson closes NC, other sites as HQ moves (http://www.google.com/hostednews/ap/article/ALeqM5gJALq9Ldcq4R3Kw55f5VqSfeNOAQD9C235SG1)." Associated Press. November 18, 2009. Retrieved on November 18, 2009.

[34] Swartz, Kristi E. " Sony Ericsson moving North American HQ to Atlanta (http://www.ajc.com/news/atlanta/sony-ericsson-moving-north-203205.html)." *The Atlanta-Journal Constitution*. Wednesday November 18, 2009. Retrieved on November 18, 2009.

[35] "Guide to Greener Electronics" (http://www.greenpeace.org/international/en/campaigns/toxics/electronics/Guide-to-Greener-Electronics/). Greenpeace International. . Retrieved 2010-08-17.

[36] "Sony Ericsson – Corporate – Company – Sustainability - Consciousdesign" (http://www.sonyericsson.com/cws/corporate/company/sustainability/consciousdesign). Sony Ericsson. . Retrieved 2010-08-17.

[37] "Ranking tables MAY 2010-Sony Ericsson.pdf" (http://www.greenpeace.org/international/Global/international/publications/toxics/2010/Ranking tables MAY 2010-Sony Ericsson.pdf). Greenpeace International. . Retrieved 2010-08-24.

External links

- Sony Ericsson Official website (http://http://www.sonyericsson.com/)

Sony Ericsson Sales Sony Ericsson Sales Figures For Quarter 2 2010 Shows a Significant decline. The Sony Ericsson phone sales volume grew over the first quarter but with a modest 4.8 percent . However, it is well up to the 2009 quarter volumes varied between 13.8 and 14.6 million units. Source (http:/ / www. letmedefine. com/ sony-ericsson-sales-figures-2010/)

Article Sources and Contributors

Sony Ericsson Xperia X10 Mini Pro *Source*: http://en.wikipedia.org/w/index.php?oldid=393737837 *Contributors*: Excirial, Fetts, Feudonym, GloomyJD, Managerarc, Soap, Svick, ZirconiumTwice, 8 anonymous edits

Sony Ericsson Xperia X10 Mini *Source*: http://en.wikipedia.org/w/index.php?oldid=398301949 *Contributors*: Bigger digger, CapnZapp, Ckatz, Dandublin93, Debnathsandeep, Destynova, Ernstp, Impact33, Jaizovic, Jt, Krazywrath, Maginks, Myscrnnm, Ocean Shores, SirMetal, Svick, 8 anonymous edits

Smartphone *Source*: http://en.wikipedia.org/w/index.php?oldid=398784292 *Contributors*: 16@r, 3Coins, A Man In Black, A123a10, A8UDI, AKA MBG, AVM, Acather96, Acolin f, Adriatikus, Aeons, Ahoerstemeier, Aillema, Akuyume, Alansohn, Alex890, Alf Boggis, AlistairMcMillan, Allen3, Amarendra.Avinash, AndrewSpec, Andries, Andros 1337, Angela, AniRaptor2001, Ansible, Arab Dynasty, Arafael, Arny, Arp120, Atama, Atenyi, Atul.ecn, Avoided, BD2412, Baronnet, Beland, Bendes, Bibijee, Biker Biker, BlkStarr, Bovineone, Branddobbe, Brianreading, Broccoli, CLW, Caj27, Calcwiki, CalumCook234, Canalteen, CatherineMunro, Changshui88, Charivari, Ciphers, Citizensmith, Cleared as filed, Closedmouth, Clovis Sangrail, Cmf, Cmlewan, Cmp101, Codetiger, Cody574, CommonsDelinker, Commontimect, Cool Hand Luke, CoolingGibbon, Cst17, Currab, Cxtom, Cyruslei, DaisyField, Dale Arnett, Daniel.finnan, DaveChild, Dcljr, Dcxf, Delphii, Deviceapps, Diego Moya, DivineAlpha, Dmarquard, Dreambringer, Drobatch, EVula, Editrrr, Elassint, Electronicguru1, Epolk, Eraserhead1, Erc, Erianna, EugeneZelenko, Eugenia loli, Evice, Evilruletheworld96, Evolutiondb, FCYTravis, Father Goose, Feinoha, Fhontoy, Flowanda, Fourdee, Furrybeagle, Fuzheado, Galadh, Garant^ ^, Geologyguy, Gilliam, Giraffedata, Glennwells, Gogo Dodo, Gomm, Gordon Ecker, GraemeL, Grahamperrin, Grajales, Graywolf, Graywolfmoon1, Greswik, Groink, Guy Harris, Gwernol, HDCase, Ha us 70, Haakon, Hadiceberg, Halloween.mac, Hamiltha, Hardylane, Harmil, Haseo9999, Hede2000, Henrik, HereToHelp, Hervegirod, Hobartimus, Hu12, Human.v2.0, Huntersquid, HuskyMoon, HybridBoy, Hydrogen Iodide, Hypnotist uk, I already forgot, I, Podius, Ianb, Ianboudreault, Ignis Fatuus, Illyria05, Immunmotbluescreen, Imroy, Imsilly123, Indefatigable, Interframe, InternetMeme, Intershark, Irky, Ishdarian, J, J.delanoy, JCDenton2052, JCrue, JNW, Jagdterrier, JamesWeb, JamieS93, Jeffq, Jerome Charles Potts, Jerryobject, Jim.henderson, JoeSmack, JonHarder, Jonabbey, Jondel, JonoP, Joseph Solis in Australia, Jsherwood0, Jsribeiro, Juan M. Gonzalez, Jusdafax, Just another editor, Jæs, Kajowi, Kariteh, KarmaGeddon, Katieh5584, Kbdank71, Kcomstock, Kellen`, KelleyCook, Kentynet, Kevin Dorner, Kingpin13, Klokbaske, KnowBuddy, Koavf, Kozuch, Kraftlos, Kresp0, Kudpung, Kuru, Kwaichi, LaVieEntiere, Lai888, Lambtron, Laurasmith70308, Lazulilasher, LeilaniLad, Lenin1991, Lester, Leujohn, Level plus, Lewispb, Lilac Soul, Limequat, Little Mountain 5, Little Professor, LrdChaos, Lukan4ica, Lun4tic, MMuzammils, Mac, Magnus.de, Malik Shabazz, Maniacgeorge, Manop, Manway, Marco.difresco, Mark Kim, Marksza, Marrowmonkey, Martarius, Masgatotkaca, Mathiastck, Mcduck, Mdrejhon, Mdwh, Meehawl, Meelar, MetaManFromTomorrow, Miaow Miaow, Mike Dillon, Mike Linksvayer, Mikehelms, Mild Bill Hiccup, Minghong, Miquonranger03, Mix Bouda-Lycaon, Mlg07e, Moberg, ModWilliam, Monaarora84, Mononomic, Morphh, MrOllie, Mvjs, MySchizoBuddy, MyronAub, Myscrnnm, Nakakapagpabagabag, Nathan94124, Navacell, Nealmcb, Neo Ogilvy, Neoarchon, Neoguy999115, Nerdeff, Netrapt, Nick Number, Night Gyr, Nightscream, Nixdorf, Nneonneo, Nopetro, Nurg, Obey, Ohnoitsjamie, Oknazevad, OlavN, Old port, Oli Filth, Omegatron, Oontay, OpenToppedBus, P.Marlow, PS., PTSE, Parintachin, Patrick, Pats1, Pbrown111, Pdahomepage, Petershank, Peyre, Phatom87, Phearson, Philip Trueman, Piano non troppo, Pingveno, Pmarshal, Pmlineditor, Pol098, Poor Yorick, Prathameshsasane, Pyfan, Querencia, Radon210, Rafael.sp, Ramu50, RedWolf, Reliablesoft, Repetition, Res2216firestar, ReverseEngineered, Reyk, Rgreed, Rhobite, Riadlem, Rich Farmbrough, Richardguk, Ron2, Ronz, SPQRobin, Sadegh87, Samad120, Samwb123, Sandstein, Sapibobo, Sdfisher, Sdrazfar, Sean13zz, Sebastian Mandrean, Serg3d2, Sfacets, Sfan00 IMG, Shalroth, Shawnc, Sheehan, Signalhead, Siliconov, Skatebiker, Skintigh, Skittle, Slitchfield, Snigbrook, Soliloquial, Spellmaster, Starofwonder, Stephen B Streater, Stephen Turner, Stjson, SuperHamster, Surenkarapetyan, Suwatest, Swoof, TVS99, Taka76, Taketa, Tangent1000, Taskinen, TastyPoutine, Tatterfly, Tfgbd, ThaWhistle, ThaddeusB, The Thing That Should Not Be, The Wild Falcon, The-apathy, Theaveng, Thestick, Thiled, Thisma, Thumperward, Ticell, Tide rolls, TimSE, Tklaer, Tmuller2, Tokyogamer, Tomas.turek18, Tombomp, Tommy2010, Tommytentimes, Toussaint, Trasz, Treekids, Tuju, UCLATre, UnrealG, Urod, Uzume, Vegaswikian, Verkinto, Verne Equinox, Versus22, Vkem, Waqas Hussain, Whkoh, WiZZLa, Wideangle, Wiki admi, Wikimhb, WikipedianYknOK, Winged-stone, Wireless Buddy, Woohookitty, XGrape, Xajel, Xrobertcmx, Youxiarock, Zanter, ZimZalaBim, ZipoB, Zpetro, 859 anonymous edits

Xperia *Source*: http://en.wikipedia.org/w/index.php?oldid=398051697 *Contributors*: AJ-India, Alexd, Anthony Appleyard, ChrisP0852, Chungxtwo, Colorred, Derek.munneke, IRTC1015, Jgp, Kajowi, Krazywrath, Ledward, Lee910, Malcolmxl5, Mernen, Mikestead, Nabil2199, Niazac, PMontanaro, Rapjul, SimonThird, Tusharpatel123, Zayani, 23 anonymous edits

Android (operating system) *Source*: http://en.wikipedia.org/w/index.php?oldid=398796766 *Contributors*: A bit iffy, A.sutton, A5b, A665321, ALIENDUDE5300, Abrahami, Aceleo, Acery, Adambiswanger1, Adamwatters, AdjustablePliers, Afriza, Aftekology, Agentlame, AladdinSE, Alboran, Alejo2083, AlexKucherenko, Alexius08, Ali'i, Alisha.4m, AlistairMcMillan, Allen Moore, Allstarecho, Alunphillips, Alvestrand, Ambictus, Ameliorate!, Ancheta Wis, Andareed, Anderssl, Andreas Carter, Andrejavus, Andresfi, Andrew Delong, Androidliscence, Andyjsmith, Anindya Bakshi, Ankitasdeveloper, AnonMoos, Anoopan, Anthonynon, Aquarat, Aradius, Arancı, Arc Orion, Arghya139, Arichnad, Arjun G. Menon, Arthas01, Aryamanjain, Aryndar, Astonmartini, Asymmetric, Athzai Khaine, Atlantia, AtteL, Attilios, Audriusa, Ayd00, Ayd000, Ayd86, Aydceri, Aydcery00, Aydcery86, Aydchery00, Aydin00, Bangbang.S, Barek, Barte, Bbaumer, Bbisgard, Bdesham, Bedna, Beland, Bender235, Berelv, Betmenko, Bevo, Binarybits, Bios Element, Blahbabe61, Blindwaves, Blowdart, Bmwtroll, Bonadea, Bosqueschool, Bovineone, Bpave777, Brandorr, Brianreading, Brianski, Briantist, Bungalowbill, C628, CCalo, CJLL Wright, CRGreathouse, Cacophony, Caltas, Can't sleep, clown will eat me, CaribDigita, Carrlos, CaseyBorders, CastAStone, Causa sui, Cbr1000f, Cburnett, Chainz, Chancer1001, ChaosData, Charles.h.white@gmail.com, CharlesC, Cherie327, Chezi-Schlaff, Chillpenguin, Chillum, Chirags, Chr1syr, Chris Bainbridge, ChrisHeller, Chrismiceli, Ciphergoth, Ciphers, ClaudeReigns, CommonsDelinker, ComputerGeek706, Conan, Conzorz, Coolbho3000, Coolstoryhansel, CoordinateFreak, Corevette, Count Chockula, Cpl Syx, Craigbrass, Csrempert, Cyberdiablo, Cyrotux, D-Notice, D20sheets, DGMDGM, DHN, DStoykov, Daabomb, Dale Arnett, Dancter, Daniel.Cardenas, Dark-Fire, Darrenm540, David Edgar, David Woodward, Davidhorman, Dawnseeker2000, Dbachmann, Dbs, Degorr, Demysc, Dennisthe2, Desbest, Dhaluza, Diamondland, Digana, Digilee, Dj.cowan, Dmit, DmitryKo, Docu, Dontmitchell, Douglaswth, Dpupkov, Dragon 280, Drbreznjev, Dreaded Walrus, Drogonov, Drrll, Dsh13, DudeThinking, Dudyk, Dueynz, Dvyjones, EEMIV, Eaefremov, Eagle-slayor, Eapache, Echeese, Ed Burnette, EdBever, Ej159, Ekerazha, Elektron, Eleman, Elronxenu, Enigmaticland, Ennustaja, Erc, Espertus, Estemi, Ettrig, Eugrus, Explorer25, Fangfufu, FatalError, Fbtjock, Feedmecereal, Ferengi, Filmore, Finalius, Fish and karate, Flatterworld, Fleminra, FlieGerFaUstMe262, Fran Rogers, Frap, FredTubale, Freddicus, Friginator, Fsamuels, Furrykef, Fxhomie, Gabriel A. Zorrilla, Gainesk, Galaxytab, Gary King, Gautamkishore, Gboxdance, Geary, Gegorg, Gh5046, Ghepeu, Ghettoblaster, Ghost650, GlassCobra, Goa103, Gogo Dodo, Gogoloid, GoingBatty, Gokberks, Golftheman, Good Olfactory, Googlemobileplatform, Gordon Ecker, GorillaWarfare, GraemeL, Grafen, Graft, Graig123, Grandscribe, GreenpeaceUbuntuMan, Gregconquest, Gronky, Gsarwa, Gsonnenf, Gu1dry, Gudeldar, Guitarguy99081, Gurch, Guru4321, Guyjohnston, Guzzyron, Gyro Copter, H4lfN3ls0n, HJ Mitchell, Haakon, Hacheema, Haggisfarm, Hanifbbz, Hcaandersen, Headinthedoor, Hedge777, Henriok, Henry W. Schmitt, Herakleitoszefesu, Hervegirod, Hgb asicwizard, Hoss789, Howlingmadhowie, Htchien, Htinlinn90, Hu12, Hucz, Hughcharlesparker, Huku-chan, Hutchinsonam, Hymek, I Feel Tired, I, Podius, I5bala, IBoy2G, IceHunter, Icep, Icydesign, Iknowyourider, IlPisano, Illegal Operation, Immunmotbluescreen, Indianstar, InternetMeme, Invenio, InvertedPendulum, Ionistii, Iridescent, Irishguy, Isaacwaller, Ivant, Ivario, J.delanoy, JAMJAM1666, JHunterJ, Jacob Poon, JacobSheehy, Jacobmathias, Jaizovic, James Foster, James Michael 1, JamesBWatson, JamesNBarnes, Jamgraham88, Jamougha, JaredMT, JavierCane, JavierMC, Jb0807, Jbreckenridge, Jcogbil, Jdthood, JeR, Jeff G., Jeffq, Jeffrey Sharkey, Jeromeds99, Jessica23, Jiess, Jimthing, Jimv1983, Jinmyo, Jmcdon10, Jmecimore, Johantheghost, JohnSawyer, Johndburger, Jokonek, Jonabbey, Jontintinjordan, Jorge Stolfi, Joriki, Josh.e.stroud, Jpvinall, Jreferee, Jrishel, Jubeidono, Julesd, Jurisnipper, Jvosika, Jwkilgore, Kaicarver, Karam.Anthony.K, Katoh, Kentyman, Kevin James Field, Kevthegreat55, Kforeman1, Khalid hassani, Kiddington, Kien64, KimDabelsteinPetersen, Kingpin13, Kiore, Kkm010, Kmdowns, Knud Winckelmann, Koavf, Kokken Tor, Komitsuki, Kontar, Korkut00, Korkut000, Kozuch, Kraftlos, Krazywrath, Krc1185, Kris cs1, Kronox android, Ksyrie, Kungming2, Kushal one, Kylelnny, LafinJack, LancerEvolution ;, Lanilsson, LarsHolmberg, LarsPensjo, Lbstone27, Legoboy920, Lenar, Lester, Lethe, Leuko, Lfcohen, Libcub, LightSpeed3, Lightenoughtotravel, Lightmouse, Lindamilton, Lindberg, Lkt1126, Llewelyn MT, LookingGlass, Lopifalko, LorenzoB, Lucas.Yamanishi, Luckerr, M0sia1, MER-C, MZMcBride, Mac, Mahanga, Male1979, Manop, Mantrik00, Mappum, Marcus Brute, Marcus2020, Mardus, Marek69, Mark Renier, Mark0528, Marko Gargenta, Marksbark, Marqueed, Martin.komunide.com, Materialscientist, Mathewsherdil, Matt Darby, Matthew0028, MatthewBurton, MattieTK, Maurice Carbonaro, Mauripop, Maximus06, McGeddon, Mcld, Meepzip, Mendaliv, Mentifisto, Messiisking, MetaManFromTomorrow, Mharen, Midgetman433, Mike.lifeguard, Mild Bill Hiccup, Millstream3, Miltonhowe, Mimihitam, Mindmatrix, Minterior, Mirabilos, Miserlou, Mistral Mktg, Mistral Solutions, Mobilepush, Modamoda, Mohanpram, Moneytoo, Moocha, Mordka, MoreThings, Mortense, Mr. Met 13, MrGALL, MrOllie, Ms2ger, Muelaner, Mugsywwiii, Mugunth Kumar, Myscrnnm, N2e, N5iln, NYKevin, Naddy, Nagy Dániel, Nahado, Namures, Nathanloop, Nealmcb, NeilN, NetHunter, Newsoxy, NexuSix, Nick Garvey, Nightscream, NiklasBr, Nikpapag, Ninly, Nogburt, Norm mit, Nuclearmoose, Nyco, Ofennell, Ohnoitsjamie, OlavN, Old Number7, Oldmokmok, Oleg Alexandrov, Oli Filth, Omshivaprakash, Orange Suede Sofa, Originalwana, OspreyPL, OwenBlacker, Oxwil, P.Shack, P2jones, PILZI, Papatenor, Pascal.Honore, Patrick, Paulscrawl, Pdfpdf, Peterkagey, Pgan002, Philip Trueman, PhosphoricX, PieterDeBruijn, Pilif12p, Pinball22, Pinecar, Pjedicke, Pkkasu, Plankhead, Plop, Pokstad, PolarYukon, Potentials, PriceChild, Prius 2, Procedure, Prolog, Pryanni, Psantora, Pseudomonas, Pvanderlee, Pwnage97, Quartermaster, Quebec99, Quoth, Qwyrxian, RScheiber, Raburton, Rajeshsweb, Ralfsmouse, Random name, Rapjul, Rapomon, Raysonho, Rborghese, RedHillian, Reedy, RegenerateThis, RenniePet, Res2216firestar, Resplendent, Rich Farmbrough, Richi, Riffic, Rigelt, Ringbang, RingtailedFox, Rmanke, Robbrown, RobertMfromLI, Roberth Edberg, Roberto.larcher, Robferrer, Rocboronat, RockMFR, Rockysmile11, Rodeosmurf, Roguegeek, RotaryAce, RoyBoy, Royce, Runtime, Rush2009, Ryan8374, RyanQuinlan, RzR, SF007, Sagarwal1981, Salamurai, Salazasu, SamJohnston, Samdman95, Sandstein, SarekOfVulcan, Sasank, Sbmeirow, Sciencewatcher, Scientus, Scl98029, Seanjacksontc, Searchmaven, SephirothXIIIX, SeyedKevin, Shachar, ShakataGaNai, SharkD, Sigma2488, Skierpage, Skudo900630, Slakr, Sleepy Sentry, Sligocki, Smashville, Smitty, Smyth, SoWhy, Solinym, Solipsys, Solomon Douglas, Soma6, Someguy1221, Speculatrix, Spiel, Sreyan, Sriram sh, StaticGull, Steel, Stefan, Stevedel7, Stevenwagner, Subbu, Suction Man, Sukael, Sunnypsyop, SvGeloven, Svetovid, Svick, Swampyank, Sylvainchevalierfu, Syndicate, Syp, T-Rex84, TMO KOTOR, TXI59, Tahitiville, Taras, TastyCakes, Tedder, Tedickey, Teeks99, The Anome, The Letter J, The359, TheEditrix2, TheTechFan, TheWishy, Thealexweb, Theanphibian, Theartfullodger, Thecurran91, Themfromspace, Thesamami, Thingg, ThomasWilson2, Thumperward, Thunder Wolf, Timeshifter, TimmmmCam, Timneu22, Tom Harris, Tomlzz1, Tomself1, Tondi5, Tony Sidaway, Torqueing, Tracer9999, TrbleClef, Trefork, Troed, Tuxcantfly, Tweisbach, UKER, UU, Ujimatcha, Ulric1313, Unamed102, Unknownwarrior33, Urashimataro, Usmanahmed25, Vadmium, Van helsing, Venona, Vincenzo.romano, Voidvector, VoluntarySlave, Vujke, WakiMiko, Walterlmitchell3, Waltonkbbl, Watchcars, Wednesday Next, Wello95, Werbej, Werdna, Wesleyarchbell, WhatMeWork, Whatiknow, Wifuk, WikiLaurent, Wikigod, Wikipedian2009, Wild mine, Williameis, Windofkeltia, Wintermute115, Wizardist, Wknight94, Wlindley, Writermonique, Wtmitchell, Wwoods, Xavierorr, Xcrivener, Xhienne, Xnamkcor, Xomm, Xrobau, Xsspider, YICbaby, Yamla, Yellowdesk, Yngvarr, Yousou, Yuriybrisk, ZacBowling, Zaratoustra, Zbutler7, Zeldex, Zero sharp, ZimZalaBim, ZirconiumTwice, Zundark, ^musaz, Ævar Arnfjörð Bjarmason, Δ, Սահակ, けいちゃ, 1495 anonymous edits

GPRS *Source*: http://en.wikipedia.org/w/index.php?oldid=15910570 *Contributors*: *drew, 24alpha, 4th guy, 63.192.137.xxx, A5b, ALM scientist, Abhinavvaid, Abldvlpr, Ahoerstemeier, Alastairgbrown, Ali@gwc.org.uk, AlistairMcMillan, Andrejj, Andros 1337, Andypdavis, Anetode, Aninhumer, Antonski, Badanedwa, Baloo rch, Banej, Barneca, Ben-Zin, Bering, Bhawani Gautam Rhk, Billylikeswikis, BirdbrainedPhoenix, Bobblewik, Cacophony, Capricorn42, Captain-n00dle, Carnildo, Carre, Charliearcuri, ChrisUK, Conti, Conversion script, Crashmatrix, Cryptext, Cvgs 0007, DaedalusRaistlin, Dafocus, Davidisom, Dawnseeker2000, Dcarriso, Defeatedfear, Depictionimage, DerHexer, Dgtsyb, DoubleBlue, Dr. Zaret, Dysprosia, EagleOne, EdoDodo, Edward, Email4mobile, EnTheMohammad, Engineerism, Ergosteur, Esnible, Europrobe, Euryalus, EwaDuan, Extraordinary, FH 3, Fleminra, Frazzydee, Fuzheado, Gaius Cornelius, Gjivan, Gzkn, Hadal, Hadiyana, Hallje, HamburgerRadio, Hede2000, Hemanshu, Hgonzale, Hhan, Hohum, Hydrargyrum, IGeMiNix, INkubusse, Iandiver, Inter, Intgr, Inzy, Jackerhack, Jacooks, Jasonauk, Jemuel, Jklin, Jnavas, Jonathanriddell, Joseph Solis in Australia, KB1KOI, Karada, KingOfSofa, Kmbsww, Ksn, Learns visits aw, Lerdsuwa, Leszek Jańczuk, LeviathinXII, Liftarn, Lightmouse, LoopZilla, Luen, MC MasterChef, MER-C, Mac, Mange01, Manumg, Mathiastck, Maximus Rex, Mentifisto, Michael Hardy, Mike Rosoft, Mittosi, Mojodaddy, Monty Dickerson, Mozzerati, N0YKG, NPalmius, Nageh, Nakon, Neale Monks, Nil Einne, Nmnogueira, Novldp, Nwynder, Ohnoitsjamie, Oli Filth, Omegatron, One half 3544, Palopt, Pan Camel, Patrick, Pb30, Pbook8989, PeteVerdon, Pgan002, PhilHibbs, Piano non troppo, Plasmaroo, Pmuschi, Pratyeka, Prolog, Qasdfdsaq, RHaworth, Radiojon, Raohammad, Requestion, Rettetast, Rich257, Richardcraib, Ricky lais, Rjstott, Roberts83, Seancdaug, Seikku Kaita, Sesu Prime, Shadowjams, Shahzad11, Shinpah1, Silsor, Silvermane, Slashme, Smalljim, Smhanov, SpaceFlight89, Starszz, Stevehughes, Stinkinrich88, Stuart Ward UK, Sudeeprg, Svinodh, Swhitehead, Tex23, The Thing That Should Not Be, Thierry Bingen, Thue, Tom Morris, Tombomp, Tommy2010, Tooki, Trevor MacInnis, TutterMouse, Utuado, Vegaswikian, Vipul, Vk2tds, Wapxana, Wik, Wikiliki, Winged-stone, Wrs1864, Wtmitchell, Youssefsan, Yury Tarasievich, Zzedar, Саша Стефановић, 454 anonymous edits

UMTS *Source*: http://en.wikipedia.org/w/index.php?oldid=15928417 *Contributors*: AAAAA, AGuerrieri, Abdull, Abune, Adamantios, Agentbla, Ale jrb, Alex.atkins, AlistairMcMillan, Amakader, Amargosa, Amelshabrawy, Andrewpayneaqa, Andros 1337, Andryono, Arkrishna, Armando82, Azurepalm, Baloo rch, Beej, BigFatBuddha, Bobblewik, Bobo192, Bovineone, Brentdax, Broccoli, Bsoft, Bwaav, Byeee, Cacophony, Cavort, CesarB, Ceyockey, Cfaerber, Cfailde, Chris Price, Chrismiceli, Christopherwoods, Contemno, Conti, Conversion script, Corpx, Crissov, Cyberglobe, Cydho, DJ Rubbie, DStoykov, DagnyB, Dale Arnett, Damian Yerrick, Dan100, Dancter, Daniel C, Danthemankhan, Darin-0, Davandron, David Johnson, DavidOrme, Dawnseeker2000, Dgtsyb, Discospinster, Dj stone, Djordjes, DmitTrix, Dog 65, DogGunn, Doris Meier, DrJolo, DrSeehas, Duhaggie, DylanW, Dziban303, Enfors, Engineerism, Epastore, Epolk, Epugachev, ErikTheBikeMan, Everyking, EwaDuan, FenSerkan, Fragglet, Frank Lofaro Jr., Frymaster, Fudoreaper, Fuzheado, GFellows, Gaius Cornelius, Gbeeker, Ghewgill, Ghoctor, Gmatsuda, Gmmour, Golbez, Goofrider, GregA, Guy Harris, Hadal, Hadiyana, Hairy Dude, Harryzilber, Henrik46, Husond, Hylaride, Imroy, InternetMeme, Intgr, JBsupreme, JPG-GR, Jackcall, Jakub, Jamelan, Jamessungjin.kim, Jasonb, Jaxl, Jaykaynam, Jdavidb, Jeffss, Jhonan, Jim.henderson, Jmgonzalez, Jmnbatista, Jmoz2989, JoeOnSunset, Joel7687, Johannylindskog, JohnTechnologist, JonHarder, Joycloete, Jpatokal, Jtact, Jupix, KGasso, KJRehberg, Karvendhan, Ke6jjj, Kennethmac2000, Kensai, KnowledgeOfSelf, Knuckleskin, Kozuch, Kphua, Kwiki, Kyng, La Pianista, Lbecque, Leion, Liftarn, Lightmouse, Lizmm, Lproven, MI canuck, MMuzammils, Maartsen, Mac, Malcohol, Mange01, Marius, Markalex, Martarius, Mathiastck, Matt croxson, Mattrix18, Maximaximax, Maximus Rex, Medconn, Michael Hardy, Michas pi, Mihhkel, Mlaffs, Modster, Mojodaddy, Monkeyblue, Mr4top, MrBeauGiles, Muhandes, MureninC, Murmurr, N0YKG, Nakedcellist, Nath85, Nedlowe, Netjeff, Nicolaiplum, Nielchiano, Nikkaro, Nil Einne, Nisselua, Nk, Noisy, NotJackhorkheimer, NotMuchToSay, Notmicro, Oli Filth, Oliver Lineham, Omegatron, Orioane, PEHowland, Pascal666, Paul Foxworthy, Paul1337, Pdelong, Petri Krohn, Picapica, Pichote, Plasticup, Porttikivi, Preslethe, Primawan, Probell, Quadratic, RAMChYLD, Rdschwarz, Rearden9, Requestion, Retroneo, Richietjpr, Richmeister, Rillian, Rjstott, Robertyhn, RoyBoy, Ruchira, Ryon, SIPH0R, Saimhe, Sajalkdas, Samuel, Sarg, Scaredpoet, Sdudah, Sendmoreinfo, Siddhant, Signalhead, SimonArlott, Skier Dude, SkylineEvo, Snickerdo, SoSaysChappy, Speedarius, Speer320, Squiggleslash, Stamppot, Stephan Leeds, Stevage, Sublite, TacoJim, Tagishsimon, TechPurism, The Anome, Thenoflyzone, Thomas Ludwig, Tide rolls, Tim Pritlove, Tobias Conradi, Towel401, Tpatricio, Travisyoung, Tsaitgaist, Uzume, Varnav, Vclaw, Vlad, Vodomar, Wamatt, Wikiliki, Wildrider99, Wk muriithi, Xp54321, YUL89YYZ, Yabbakazoo, Youssefsan, Yubal, Zigger, Zoicon5, Zquack, 551 anonymous edits

HSDPA *Source*: http://en.wikipedia.org/w/index.php?oldid=16402882 *Contributors*: 1wonjae, Abbasah, Abrech, Abune, AdeMiami, Adhame95, Admrboltz, Agentbla, Alfpooh, AlisonW, Analogue Kid, Andrewmp, Andros 1337, Anetode, Angelsfreeek, Anritsu2007, Ansend, Arado, Armando82, AspiringEntrepreneur, BTTNext, Bender235, Bigbmc26, Bingobangobongoboo, Blathering1, Bobblewik, BokicaK, Bonnie116, Brambi, Bryan Derksen, Bwaav, Bz2, C0nanPayne, CalumH93, Camw, Canberra photographer, Capricorn42, CezarkennySeF, Chancheelam, CharlBarnard, Chids15, Chockyboy, Chrisbolt, Chulk90, Cipherswarm, Clesch, Cmcfarland, Colonies Chris, Cometstyles, DH85868993, DKEdwards, Dale Arnett, Darin-0, Darkov, David Edgar, David Legrand, DavidWBrooks, Dawnseeker2000, Dcoetzee, Debresser, Deville, Dicksonfu, Didipg, Diego-en, Dilydan, DocWatson42, Dodeeric, Dog 65, Drizzd, Drr797, DylanW, Ed Yeon, Editore99, Edolen1, EinarKramer, Engineerism, Ernstp, Eskalin, Evdo, Evilspoons, EwaDuan, Family Guy Guy, Ferdinand Pienaar, Feureau, Fiftyquid, Firsfron, Fr33Spac3, Fred Bradstadt, Frenchwhale, Gaius Cornelius, Galaxiaad, Ghalas, Gsarwa, Gudjon, Gurch, Guteq, Hadj, HeffeQue, HoodedMan, Hoosss, Howardchu, IRP, Iceberg3k, Intgr, Isilanes, Isnow, JForget, Jamsta, Jfinlayson, Jhdaly, Jinhuili, Jmalonzo, Jonohill, JoshiFarron, Jverlin, Kenguest, Khalid hassani, Kiki47, Kjkolb, KnightRider, Kozuch, Kraftlos, Kusunose, Kyng, LarsHolmberg, Lesteague, Lightmouse, Lmendo, Lordfani, Lukemcurley, MMuzammils, Mac, Mafmafmaf, Mange01, Mariodrs, Mathews sunny, Mathiastck, Matt croxson, Mauls, Maxhaase, Megaloman, Mlorimer, Mojodaddy, Mroach, Muzeri, N3c, Nagle, NeilenMarais, Neor0001, Neutron939, Nick, Nikolaj Christensen, Nk, Nneonneo, Nukeless, Nzlander, Oayk, Occuli, Oli Filth, Ollihokkanen, P a morrison, PanMan, Panscient, PauliKL, Paxvobiscum, Pepijn Schmitz, PeterZacSmith, Pgan002, Polkapunk, Psarmstr, RTG, Radiojon, Rajqo, Rbirkby, Remuel, Requestion, Retroneo, Rewt, Rich Farmbrough, Rjwilmsi, Rlove, Robinw77, Robwingfield, Ronline, Rspanton, RyanCu, Ryuu, Sajaji, Sathish.visu, Sesu Prime, Sfoskett, Sharcho, Shrirang, Siddhant, Silvestre Zabala, Simon hibbs, Simongnz, Skarebo, Slgcat, Smitthy, Smnc, Smoothy, Snickerdo, Solipsist, Soumyakanti, Squiggleslash, Sranjanm2002, Statiklicious, SteinbDJ, Sublite, Svick, Szlevi, Template namespace initialisation script, TerraFrost, Tevulytis, Tex23, TigerK 69, Timsdad, Tntuof, Tonis1, Towel401, Twinxor, Txuspe, Vec, Vegaswikian, Vlad, Wibbble, Wiki alf, Winterspan, WirelessMoves, Wirelessman, Wwmbes, Xangel, YordanGeorgiev, Zeeraha, Zilog Jones, Zobisch, Zollerriia, Zr40, 639 anonymous edits

Sony Ericsson XPERIA X10 Mini Pro *Source*: http://en.wikipedia.org/w/index.php?oldid=377219315 *Contributors*: Excirial, Fetts, Feudonym, GloomyJD, Managerarc, Soap, Svick, ZirconiumTwice, 8 anonymous edits

Bluetooth *Source*: http://en.wikipedia.org/w/index.php?oldid=398389375 *Contributors*: 10metreh, 1wolfblake, 417.417, 5 albert square, 6birc, AEMoreira042281, AKMask, ARUNKUMAR P.R, Aapo Laitinen, Aaronsharpe, Abesford, Abidh786, Abomasnow, Acer65, Adashiel, Addea, AdjustShift, Adovid-Mila, Adrianfleming, Aeon1006, Aezram, AfrowJoww, Agency SEA, Aguwhite, Ahmadr, Ahruman, Ain't love grand, Airplaneman, Aitias, Ajbrowe, Aka042, Alansohn, Alella, Alex.g, Alexey V. Molchanov, Alexf, Alias Flood, AlistairMcMillan, Alphachimp, Altenmann, Altermike, Alyssa3467, Amigadave, Anderssjosvard, Andrea105, Andreasbecker, AndrewRH, Andrewmc123, Andy120, Andypandy.UK, Angela, Angrymuth, Animum, Anon lynx, Anonymi, Antandrus, Aphid360, Apparition11, Aprilstarr, Apterygial, Apyule, Aquinex, Arabani, ArglebargleIV, Arienh4, Arjun G. Menon, Armando82, Arthur Rubin, AubreyEllenShomo, Auric, Ausinha, Austin Spafford, Avaldesm, Avalyn, Averisk, Avijaikumar, Avoided, AxelBoldt, Aymatth2, B, BRUTE, Band B, Bardia52, Bdoserror, Beland, Bfigura's puppy, Bigjimr, BingoDingo, Bitbit, Bk0, Blehfu, Blob00, Bloodofox, Blu3tooth, BlueMint, BobKawanaka, Bobblewik, Bobo192, Boffob, Bogey97, Bongwarrior, BonzoESC, Boomshadow, Bovineboy2008, Bpantalone, BradBeattie, Brahmanknight, Brent01, Brianski, BrockF5, Brougham96, Bruns, Bsadowski1, Btornado, Bucketsofg, Burchseymour, Butter Bandit, CQJ, CWii, Caltas, Cambrant, Can't sleep, clown will eat me, Canadian-Bacon, Canadiantire122, Canterbury Tail, Cape ryano, Capricorn42, Carradee, Casimir, CasperGhosty, Catgut, Cathcart, Cburnett, Ceoil, Chalo phenomenon, Charles Gaudette, CharlieGalik, Cheeseguypie, Chillum, Cmichael, Colfer2, Color probe, ComCat, Compellingelegance, Conversion script, CoolFox, Coolbho3000, CosminManole, Cowpriest2, Cpl Syx, Crakkpot, Cransdell, CrazyTerabyte, Crf150r rider, Crissov, Curps, D1991jb, DKBG, DSRH, DV8 2XL, Da monster under your bed, Dacbook, Dadude3320, Daemonic Kangaroo, Dajhorn, Danfuzz, Danno uk, Dannydream9999, Darguz Parsilvan, Darin-0, Darth Panda, Darz Mol, Dat789, Dave Andrew, Dave4400, Davegsun, Daveswagon, David Haslam, David Martland, David Parker, DavidCary, DavidConrad, Davidbspalding, Davidgothberg, Davidoff, DavisLee, Davrukin, Del Merritt, Delirium, Demian12358, Demonte Morton, Dennis mathews, Depaul27, DerHexer, Deskana, Dicklyon, DigitHead, Digitallyinspired, Dilane, Dimosvki, Dinomite, DireWolf, Dirkbb, Disavian, Discospinster, Dj4u, Djg2006, Dlohcierekim, Dmytro, Docboat, Doggkruse, Doniago, DopefishJustin, Dorvaq, DoubleBlue, DougsTech, Dpotter, Dpwkbw, Dracony, Dragon 280, Drano, Drbreznjev, Droll, Droob, Dtgriscom, Dtt.hatchi, Duffman, Dumelow, Dusti, Dwaipayanc, Dwlocks, Dwo, Dyfrgi, Dysprosia, Dystopos, Dzero-net, EJF, EVula, Ears04, Ed Poor, Ed g2s, Edcolins, Edward, EdwinHJ, Efthimios, Egwyn3, Ehn, Eiffel, Eisnel, El C, Electron9, Elopezar, Enap007, EncMstr, Energetic is francine@yahoo.com, Engineerism, Enviroboy, Epbr123, Eptalon, Eptin, Ericg, Eskalin, Everyking, Ex nihil, Excirial, Ezra Wax, FCYTravis, FT2, Face, Faisal.akeel, FakeTango, Falcon8765, Farosdaughter, Fartacus46, Fatespeaks, Favonian, Fdlj, Felixgomez18, Felyza, Femto, Fieldday-sunday, Fireblaster lyz, Flatline, Flcelloguy, Foobar, Frazzydee, Freakofnurture, Freitasm, Friviere, Froid, Frvwfr2, Funkendub, Funnyfarmofdoom, Fxhomie, G-W, GFellows, Gadfium, Gadgetmonster, Gaius Cornelius, Galaxiaad, Garion96, Gary63, Genia4, Geoman888, Ghen, Giftlite, Gilliam, Ginettos, Glarrg, Glenn, Glenn4pr, Gloriamarie, Gmoose1, Gobeirne, Gogo Dodo, Golbez, Gophi, Gracenotes, GraemeL, Grafen, Graham87, Granpire Viking Man, Greensalad, Greenshed, Griffin5, Gringer, Gromreaper, Grungen, Grunt, Gscshoyru, Guanaco, Gustavopaz21, Gwernol, Haakon, Hadal, Hahnchen, HalfShadow, HannesP, HappyCamper, Happysailor, HarisM, Harleyosborne1234, Harvester, Haseo9999, Haukurth, Hdante, Headsetuser, Heimstern, Hellisp, Heman, Herbythyme, Heron, Hersfold, Hfkids, Hitchcockc, Hk1992, Holdenmcrutch, Holt, Horkana, Hoseabrown, Howie1989, Hu12, Huddsy09, Hyst, I already forgot, IMSoP, Iam, Iammasud, IanGM, Icestorm815, IddoGenuth, Ifroggie, Igoldste, Impasse, Impi, Imroy, Imz, Infrogmation, Intgr, Iridescent, Irstu, Isidore, Iterator12n, Ixfd64, J.delanoy, JFreeman, JKofoed, JLaTondre, JRaue, JSimmonz, JSpung, Jakro64, James Michael 1, JamesBWatson, Jao, Jas203, Jasonataylor, Jasonauk, Jaysbro, Jblbt, Jdlh, Jebba, Jeffq, Jench, Jennypei, Jerec, JeroenvW, JerryDeSanto, Jfinlayson, Jfromcanada, Jgrahn, Jhsounds, Jj137, Jlarkin, Jlou us, Jmnbatista, Jnavas, JoanneB, Joconnor, JoeMarfice, Joey7643, John Anderson, John Quincy Adding Machine, John Reaves, John254, John2kx, JohnOwens, JonHarder, Jonathan Hall, Jonathan Karlsson, Jonkerz, Jonverve, Jossi, Jsmaye, Jtmoon, Julesd, Jusdafax, Justallofthem, Juxo, Jw-wiki, Jwissick, Jóna Þórunn, Kaini, Kamasutra, Karichisholm, Karl2620, Katimawan2005, Katoh, Kazu89, Keilana, Kelaiel, KelleyCook, KenBentubo, Kenb215, Kenguest, Kerr avon, Kevin chen2003, Khalid hassani, Kierenj, Killdevil, Kimiko, Kirrages, Klaus100, Klensed, Koman90, Kowloonese, Kozuch, Kralizec!, Kravdraa Ulb, Ksnow, Ktims, Kungfuadam, Kuru, Kurykh, Kyng, LA2, La Pianista, Lakers, Landon1980, Landroo, Laspalmas, Laurusnobilis, Lavenderbunny, LeadSongDog, Learnportuguese, LeaveSleaves, Leibniz, Leif, Leujohn, Leuko, Lexischemen, Lexor, Liftarn, Lightmouse, LilHelpa, Lilboogie, Lipatden, Lisa 4 envyme@yahoo.com, Little Mountain 5, LittleOldMe, Littleworld73, Live Forever, Lohith92, LokiClock, Lonelydarksky, Lordkazan, Loren.wilton, Lotje, Lt Master1, Luccas, Luk, Lupo, Lwc, Lzur, MAlvis, MER-C, MG IUkE xX, Mabahj, Mac, Macy, Mahjongg, Makeemlighter, Mani1, Manticore, Marcelbutucea, MarkS, Markus Kuhn, Martinp23, Masamage, Masatran, Masgatotkaca, Matt Crypto, Maurice45, Maxamegalon2000, Maximaximax, Mckinleyma, Mcsee, Mdwyer, Meckleys, Melaen, Melnorme, Melsaran, Mendel, Mercury, Merlion444, Mholland, Michael Hardy, MichaelKirche, MichaelStanford, Michkalas, Mike Rosoft, Millionbaker, Minna Sora no Shita, Mipadi, Misterdan, Misza13, Mithaca, Mjmarcus, Mltinus, Mntlnrg, MobileMistress, Modster, Mojtaba fouladi, Mooquackwooftweetmeow, Morte, Morten, Morwen, Mossig, Mowgli, Mpesce, MrOllie, MrPaul84, MrRedwood, Mrand, Mrbluetooth, Mrmason, Mschlindwein, Mtodorov 69, Mushroom, Mushu118, Mwanner, Mwilso24, Myscrnnm, Mythril, N.o.bouvin, N2e, NAHID, NHRHS2010, NMHartman, NXTguru, Nachoman-au, Nagle, Nagy, Nakon, Narendra Sisodiya, Natalya, Natgoo, Navinkhemka, NawlinWiki, Nburden, Ndenison, Nentuaby, Neon white, NetRollller 3D, Netesq, Nettlehead, Neutrality, Newtruie, Nfearnley, Nialsh, Nichalp, NickW557, Nigholith, Nikai, Nikhildandekar, Nile, Ninja Wizard, Nirvana2013, Nisanthks, Nixdorf, Nlu, Nobunaga24,

Noctibus, Nopetro, Notbyworks, Notmicro, Novasource, Nux, Nylex, O mores, Oden, OekelWm, Ohnoitsjamie, OlEnglish, OlavN, Olejarde, Oli Filth, Olorin28, Omegatron, Omicronpersei8, Omnilord, Oo7565, Opelio, Ouishoebean, Oxymoron83, P Carn, P3net, PEZ, PRRfan, PS2pcGAMER, Padishar, Palopt, Parsecboy, Patcat88, Paul Richter, Pavel Vozenilek, Pedess, Pedrose, Persian Poet Gal, Peruvianllama, Peter M Gerdes, Peyre, Pfaff9, Phantomsteve, Pharaoh of the Wizards, Phil Boswell, Philip Trueman, Philip123ison, Phoenix-forgotten, PhoenixofMT, Phooto, Piano non troppo, Pigsonthewing, Pikemaster18, Plagana, Plasticbadge, PleaseStand, Plexos, Pmaas, Pmsyyz, Pogo747, Pol098, PolarYukon, Polo200, Porqin, Poterxu, ProbablyDrew, Proximitia, Przemyslaw Pawelczak, Psmonu, Pstanton, Puckly, PuerExMachina, Pw201, QrK FIN, Quartz25, Queenmomcat, Qwavel, Qxz, RJaguar3, Rack1600, Raistolo, Rakeshlashkari, Ramu50, Random Nonsense, Recognizance, RedWolf, Rediahs, Repetition, Retiono Virginian, Retired username, Retroneo, RexNL, Rfc1394, Rhallanger, Rich Farmbrough, Rob Blanco, Robert K S, RobyWayne, Rocastelo, RodC, Romanm, Ronnymexico, Ronz, Roybadami, Rufous, Runmcp, RyanCross, Ryanbrannonrulez, S3000, SJP, SLi, SMC89, SQB, ST47, Salty-horse, SammyJames, SampsonSimpson, SandyGeorgia, Sannse, Sappe, Sapphic, Sars, Sceptre, SchfiftyThree, Schwarzes Nacht, Sciurinæ, Scott Wilson, Scott14, Sdream93, Seidenstud, Senator Palpatine, Servel333, Seth Nimbosa, Sfoskett, Shaddack, Shadowjams, Shadowlink1014, Shadowlynk, Shahad, Sharkface217, SheldonYoung, Shoeofdeath, Shyhshinlee, Siddhant, Sietse Snel, Sillybilly, Sionus, Sjakkalle, Sjjupadhyay, Skarebo, Skyezx, Skyskraper, Skyy Train, Slakr, SleepyHappyDoc, Sleske, Snoopy8765, SoCalSuperEagle, Somearemoreequal, Someguy1221, Someone else 90, Sonic3KMaster, Sonu27, SpK, Spab 007, Spaceman85, Spellcast, Splash, Squeakypaul, Squid tamer, Squirrelist, Ss power hacker, Ssd, Ssri1983, StaticGull, Ste m t1988, Stephan Leeds, Stephen B Streater, Stephenb, Steven Zhang, Steveobd, Sthow, Stuffed cat, Sudsak, Sunray, Supernovaess, Suruena, Synchronism, Szumyk, THA-Zp, THEN WHO WAS PHONE?, TPK, Tabledhote, Talkie tim, Tanvir Ahmmed, Tarret, Tauheedrameez, Th1rt3en, The Epopt, The Flying Spaghetti Monster, The Phoenix, The Rambling Man, The Thing That Should Not Be, The undertow, TheBilly, TheDoctor10, TheOtherSiguy, Thedjatclubrock, Theflyer, Theups, Thingg, Thinkmike, Tide rolls, TigerShark, Tim1357, Timo Honkasalo, Tkgd2007, Tmopkisn, Tobias Bergemann, Todd Vierling, Todeswalzer, Tokachu, Tompagenet, Tomvanbraeckel, Tony1, Tooki, Toolnut, ToothTeam, Topazg, Towel401, Tpbradbury, Travelbird, Trent Arms, Tresiden, Trevor MacInnis, Tris121, TutterMouse, U235, Uker, Uncle Dick, Uncle G, Unschool, Utcursch, V 2e, Vadmium, VegaDark, Vegaswikian, Versus22, Vid2vid, Village Idiort, Vina, Vinod58, Virexix, Vkem, Vojamisic, WMC, WadeSimMiser, Wahooker, Warpedshadow, Warreed, Wdrdoctor, Wehe, Weihao.chiu, Werdan7, WesleyDodds, Westleyd, Whosyourjudas, Whpq, WibblyLeMoende, WickedInk, Wik, WikHead, WikiLaurent, WikiMarcus, Wikiborg, Wikieditor06, Wikitumnus, William M. Connolley, William Wang, Willkn, Wimt, Winchelsea, Winesouvenir, Wireless friend, Wisq, Wk muriithi, Wkdewey, Woohookitty, WookieInHeat, WriterHound, Wtmitchell, Wtshymanski, Wvbailey, Ww, Xelgen, Xenium, Xp54321, Xpclient, Y2kcrazyjoker4, Yahia.barie, Yakudza, Yamamoto Ichiro, YhnMzw, Yidisheryid, Yyy, Zebov, Zhou Yu, Zidane2k1, Zlatan8621, Zpetro, Zsinj, Åkebråke, 2383 anonymous edits

Sony Ericsson *Source*: http://en.wikipedia.org/w/index.php?oldid=398521352 *Contributors*: -Majestic-, A.h. king, A.szczep, A4, ADouBTor, AL3X TH3 GR8, Abijwe, Adamo117, Ageo020, Agravius, Ahoerstemeier, Alexaq, Alexd, Alfio, AlfredWalsh, AlistairMcMillan, Alpha 4615, Aman.mutreja, Amitsingh71, Andrenamarkley, Andros 1337, Anger2headshot, Anna Lincoln, Anthonygrig, ArchonMagnus, Arsenikk, Avala, Barefootguru, Bearcat, Bilbo571, Bilky asko, Blhilbrands, Bongomatic, Butterfly0fdoom, Capricorn42, Carbonrodney, CaribDigita, Carina22, Carlsotr, CezarkennySeF, Charea, Chris the speller, Christopher Mahan, Ckatz, Clipmode, Closedmouth, Cmichael, Collinhogorman, Colonies Chris, Cometstyles, CommonsDelinker, Copie0178, Crysb, Cynical, Cyrius, DARTH SIDIOUS 2, Dahlis, Dan100, Dancter, Daniel.Cardenas, Danielphin, DarkSaber2k, David0811, Davidswan, Deguld, DerHexer, Desplow, Discospinster, Djr xi, Donovan1983, Dostick, DoubleBlue, Dpol, Dr.frog, Duffman, Duk, Dysprosia, Ecowiki, Eliethesame, Elliot Richards, Emersoni, EnTheMohammad, Enemenemu, Epbr123, Ericselga, Erwe12, EvocativeIntrigue, Exity, Ezeu, Firsfron, Folksong, Ftpaddict, G4nt1, Gail, GoldenGoose100, GreenJoe, GregorB, H3llbringer, Haakon, HaeB, Halsteadk, Harcalion, HarryMichaelJones, Hatesonyericsson, Hayabusa future, Hemanshu, HenryLi, Hill jsr, Horsee, Huaiwei, IW.HG, Iceduck, Ief, Igno2, Imroy, Infogeeks, Iridescent, Iridium ionizer, Irishguy, Irstu, Ixfd64, J Di, J.delanoy, JaGa, Jamcib, Jaraalbe, Jdtyler, JeremyA, Jerryseinfeld, JessaRinaldi, Jiggerwiggles, Jimfbleak, Jjkope, Jm51, Jmlk17, Johnnydontlook, Jon Paul Janet, Jontintinjordan, Jummplaum, Justmejeff, KaozGamer, Katous1978, Kkm010, Klamsd, Kocio, Konieckropka, Kxdan, LOL, Lakshmix, Larry laptop, Laxstar5, Lensovet, Leolaursen, Lepetitvagabond, Lightlowemon, Looie496, Lovely Chris, LudwigVonBirkenstock, Lzur, M3lm4tt, MER-C, Mabdul, Macauleyd4, Magnum2037, Mark83, Martarius, Martnym, Master of Puppets, Matthew Yeager, Maxhaase, Maxis ftw, Mdiasif, Megahmad, Mephistophelian, Mic, MidgleyDJ, Milan292208, Mirmo!, Misza13, Mitsukai, Monkeyman, Moocowsrock, Mrt396, Mx3, Mygerardromance, Myscrnnm, NHRHS2010, Nagy, Nateji77, Neo-Jay, Netrat, Niall123456789, NightFalcon90909, Nikola Smolenski, Nitya Dharma, Nneonneo, NuclearTourist, OOODDD, Ocrho, Ojay123, Oli Filth, Opelzafira, Orgtelecom, Ostralek, PTSE, Pappfer, Patstuart, Pavel Vozenilek, Pencilcase123, Phantomsteve, Phonefinder2007, Piano non troppo, Pip2andahalf, Plop, Postdlf, Prolog, Psha, Pudeo, Puredesi123456789, Quintote, RTG, RadioActive, Rafpilotdavid, RainbowOfLight, Raviaka Ruslan, Ravichandar84, Rcaelius, Rcandelori, Redsaph, Reeloo, Relata refero, Repetition, Ricardocolombia, Rigelt, Rimmington01, Rjransijn, Rjwilmsi, Roguegeek, Ronz, Rupertslander, S.K., SLm4ever, Sabri76, Salgueiro, Sanmint, Scotsboyuk, Sean2012, Secaundis, Semobilephones, Serlin, Shaliron, Shandris, Shell Kinney, Shipmaster, Siliconov, SimonMenashy, Sir Edgar, Sirimiri, Sjakkalle, Sl, Slavon37, Smadge1, Smarites-, Smyth, Sohailstyle, SpaceFlight89, Speedboy Salesman, SpigotMap, Srushe, Ssolbergj, Stephantom, Sunholm, T, TMC1982, Takamaxa, Tango, Tavdy79, Tdu1, Tedats, The Negotiator, Theda, Themfromspace, Tide rolls, TigerK 69, Tresiden, TriangleBelow, Tutlearner, UkPaolo, Ukeu, Ulric1313, Unintentional Guy, V lochan89, Varundbest10, Vegas Bleeds Neon, Vegaswikian, Vespristiano, Viakenny, Vikingstad, Visik, W2ch00, WeisheitSuchen, WereSpielChequers, WhisperToMe, Whitepaw, Wibbble, WichitaQ, Wickethewok, Wiki Wonka 2.0, Wiki0709, Wikipelli, Woohookitty, Workingbigtime, Wtni, Yakiv Gluck, Yas, Yeeshenhao, Youngjoon Shin, Yrulee, Yzmo, ZippySLC, Zpetro, 692 anonymous edits

Image Sources, Licenses and Contributors

File:Smartphone_share_current.png *Source*: http://en.wikipedia.org/w/index.php?title=File:Smartphone_share_current.png *License*: GNU Free Documentation License *Contributors*: -- Eraserhead1 <talk> 12:49, 3 March 2010 (UTC) Graph created by myself. Original uploader was Eraserhead1 at en.wikipedia

File:Android logo.svg *Source*: http://en.wikipedia.org/w/index.php?title=File:Android_logo.svg *License*: Attribution *Contributors*: Google and Android's developpers

File:Android-2.2.png *Source*: http://en.wikipedia.org/w/index.php?title=File:Android-2.2.png *License*: unknown *Contributors*: Android Developers http://developer.android.com/index.html

Image:Android home.png *Source*: http://en.wikipedia.org/w/index.php?title=File:Android_home.png *License*: GNU General Public License *Contributors*: Unamed102

File:Diagram android.png *Source*: http://en.wikipedia.org/w/index.php?title=File:Diagram_android.png *License*: GNU Free Documentation License *Contributors*: Alvaro Fuentes Vasquez (Kronox)

Image:Android mobile phone platform early device.jpg *Source*: http://en.wikipedia.org/w/index.php?title=File:Android_mobile_phone_platform_early_device.jpg *License*: Creative Commons Attribution 2.0 *Contributors*: Kai Hendry from London, UK

File:Android-logo.jpg *Source*: http://en.wikipedia.org/w/index.php?title=File:Android-logo.jpg *License*: Attribution *Contributors*: Google and Android's developpers

Image:Android.svg *Source*: http://en.wikipedia.org/w/index.php?title=File:Android.svg *License*: unknown *Contributors*: Binnette, Rocket000, WikipediaMaster

Image:Android os distribution oct 10.png *Source*: http://en.wikipedia.org/w/index.php?title=File:Android_os_distribution_oct_10.png *License*: Creative Commons Attribution 2.5 *Contributors*: Android Open Source project

Image:Huawei E220 (Three).jpg *Source*: http://en.wikipedia.org/w/index.php?title=File:Huawei_E220_(Three).jpg *License*: Creative Commons Attribution 3.0 *Contributors*: User:Korax1214

File:UMTS Network Architecture.svg *Source*: http://en.wikipedia.org/w/index.php?title=File:UMTS_Network_Architecture.svg *License*: Creative Commons Attribution-Sharealike 3.0 *Contributors*: User:Tsaitgaist

File:UMTS-fridge.jpg *Source*: http://en.wikipedia.org/w/index.php?title=File:UMTS-fridge.jpg *License*: GNU Free Documentation License *Contributors*: User:PPP

File:Option GT 3G+ UMTS card.jpg *Source*: http://en.wikipedia.org/w/index.php?title=File:Option_GT_3G+_UMTS_card.jpg *License*: Creative Commons Attribution-Sharealike 3.0 *Contributors*: User:Woookie

Image:Nokia6650 unlocked.jpg *Source*: http://en.wikipedia.org/w/index.php?title=File:Nokia6650_unlocked.jpg *License*: Public Domain *Contributors*: User:Towel401

Image:Bluetooth.svg *Source*: http://en.wikipedia.org/w/index.php?title=File:Bluetooth.svg *License*: unknown *Contributors*: 16@r, AEMoreira042281, Alex rosenberg35, Anomie, BrockF5, MBisanz, Neo139, Plasticspork, 7 anonymous edits

Image:H-rune.gif *Source*: http://en.wikipedia.org/w/index.php?title=File:H-rune.gif *License*: Public Domain *Contributors*: Dbachmann, Holt, Maksim, 2 anonymous edits

Image:Runic letter berkanan.svg *Source*: http://en.wikipedia.org/w/index.php?title=File:Runic_letter_berkanan.svg *License*: Public Domain *Contributors*: User:ClaesWallin

Image:Product1.jpg *Source*: http://en.wikipedia.org/w/index.php?title=File:Product1.jpg *License*: Free Art License *Contributors*: AllProducts.com

Image:Drone 4.jpg *Source*: http://en.wikipedia.org/w/index.php?title=File:Drone_4.jpg *License*: Public Domain *Contributors*: User:Mmckinley

Image:BluetoothUSB.jpg *Source*: http://en.wikipedia.org/w/index.php?title=File:BluetoothUSB.jpg *License*: Public Domain *Contributors*: Original uploader was Abidh786 at en.wikipedia

Image:DELL TrueMobile 350 Bluetooth card.jpg *Source*: http://en.wikipedia.org/w/index.php?title=File:DELL_TrueMobile_350_Bluetooth_card.jpg *License*: unknown *Contributors*: User:Omegatron

File:Sony Ericsson logo.svg *Source*: http://en.wikipedia.org/w/index.php?title=File:Sony_Ericsson_logo.svg *License*: unknown *Contributors*: -Majestic-, Alx 91, PhilKnight, Rjd0060, Wiikipedian, 4 anonymous edits

File:Decrease2.svg *Source*: http://en.wikipedia.org/w/index.php?title=File:Decrease2.svg *License*: Public Domain *Contributors*: User:Sarang

Image:SonyEricsson income2003to2009.png *Source*: http://en.wikipedia.org/w/index.php?title=File:SonyEricsson_income2003to2009.png *License*: Creative Commons Attribution-Sharealike 3.0 *Contributors*: User:Enemenemu

File:SE K750i.jpg *Source*: http://en.wikipedia.org/w/index.php?title=File:SE_K750i.jpg *License*: Creative Commons Attribution 2.0 *Contributors*: User:Martin Seyer

Image:SonyEricsson shipments2003to2009.png *Source*: http://en.wikipedia.org/w/index.php?title=File:SonyEricsson_shipments2003to2009.png *License*: Creative Commons Attribution-Sharealike 3.0 *Contributors*: User:Enemenemu

License

GNU Free Documentation License Version 1.2, November 2002 Copyright (C) 2000,2001,2002 Free Software Foundation, Inc. 59 Temple Place, Suite 330, Boston, MA 02111 -1307 USA Everyone is permitted to copy and distribute verbatim copies of this license documen t, but changing it is not allowed.

). PREAMBLE

The purpose of this License is to make a manual, textbook, or other functional and useful document "free" in the sense of reedom: to assure everyone the effective freedom to copy and edistribute it, with or without modifying it, either commercially or noncommercially. Secondarily, this License preserves for the author and publisher a way to get credit for their work, while not being considered responsible for modifications made by others. This License is a kind of "copyleft", which means that derivative works of the document must themselves be free in the same sense. It complements the GNU General Public License, which is a copyleft license designed for free software. We have designed his License in order to use it for manuals for free software, because free software needs free documentation: a free program hould come with manuals providing the same freedoms that the oftware does. But this License is not limited to software manuals; can be used for any textual work, regardless of subject matter r whether it is published as a printed book. We recommend this icense principally for works whose purpose is instruction or eference.

. APPLICABILITY AND DEFINITIONS

This License applies to any manual or ot her work, in any nedium, that contains a notice placed by the copyright holder aying it can be distributed under the terms of this License. Such notice grants a world -wide, royalty -free license, unlimited in uration, to use that work under the conditio ns stated herein. The Document", below, refers to any such manual or work. Any nember of the public is a licensee, and is addressed as "you". ou accept the license if you copy, modify or distribute the work a way requiring permission under copyright l aw. A "Modified ersion" of the Document means any work containing the ocument or a portion of it, either copied verbatim, or with nodifications and/or translated into another language. A Secondary Section" is a named appendix or a front -matter ection o f the Document that deals exclusively with the elationship of the publishers or authors of the Document to the ocument's overall subject (or to related matters) and contains othing that could fall directly within that overall subject. (Thus, if ne Docu ment is in part a textbook of mathematics, a Secondary ection may not explain any mathematics.) The relationship could e a matter of historical connection with the subject or with elated matters, or of legal, commercial, philosophical, ethical or olitical position regarding them. The "Invariant Sections" are ertain Secondary Sections whose titles are designated, as being ose of Invariant Sections, in the notice that says that the ocument is released under this License. If a section does not fit e above definition of Secondary then it is not allowed to be esignated as Invariant. The Document may contain zero variant Sections. If the Document does not identify any Invariant ections then there are none. The "Cover Texts" are certain short assages of text that are listed, as Front -Cover Texts or Back - over Texts, in the notice that says that the Document is leased under this License. A Front -Cover Text may be at most words, and a Back -Cover Text may be at most 25 words. A ransparent" copy of the Document means a machine -readable py, represented in a format whose specification is available to e general public, that is suitable for revising the document raightforwardly with generic text editors or (for images omposed of pixels) generic p aint programs or (for drawings) me widely available drawing editor, and that is suitable for input text formatters or for automatic translation to a variety of rmats suitable for input to text formatters. A copy made in an herwise Transparent file format whose markup, or absence of arkup, has been arranged to thwart or discourage subsequent odification by readers is not Transparent. An image format is t Transparent if used for any substantial amount of text. A copy at is not "Transparent" is called "Opaque". Examples of suitable rmats for Transparent copies include plain ASCII without arkup, Texinfo input format, LaTeX input format, SGML or XML ing a publicly available DTD, and standard -conforming simple ML, PostScript or PDF designed f or human modification. amples of transparent image formats include PNG, XCF and G. Opaque formats include proprietary formats that can be ad and edited only by proprietary word processors, SGML or L for which the DTD and/or processing tools are not generally ailable, and the machine -generated HTML, PostScript or PDF oduced by some word processors for output purposes only. The tle Page" means, for a printed book, the title page itself, plus ch following pages as are needed to hold, legibly, the material s License requires to appear in the title page. For works in mats which do not have any title page as such, "Title Page" ans the text near the most prominent appearance of the work's e, preceding the beginning of the body of the te xt. A section ntitled XYZ" means a named subunit of the Document whose e either is precisely XYZ or contains XYZ in parentheses owing text that translates XYZ in another language. (Here XYZ nds for a specific section name mentioned below, such as cknowledgements", "Dedications", "Endorsements", or story".) To "Preserve the Title" of such a section when you dify the Document means that it remains a section "Entitled Z" according to this definition. The Document may include rranty Disc laimers next to the notice which states that this ense applies to the Document. These Warranty Disclaimers considered to be included by reference in this License, but y as regards disclaiming warranties: any other implication that se Warranty Disclaimers may have is void and has no effect the meaning of this License.

VERBATIM COPYING

u may copy and distribute the Document in any medium, er commercially or noncommercially, provided that this ense, the copyright notices, and the license notice saying this License applies to the Document are reproduced in all copies, and that you add no other conditions whatsoever to those of this License. You may not use technical measures to obstruct or control the reading or further copying of t he copies you make or distribute. However, you may accept compensation in exchange for copies. If you distribute a large enough number of copies you must also follow the conditions in section 3. You may also lend copies, under the same conditions stated ab ove, and you may publicly display copies.

3. COPYING IN QUANTITY

If you publish printed copies (or copies in media that commonly have printed covers) of the Document, numbering more than 100, and the Document's license notice requires Cover Texts, you must enclose the copies in covers that carry, clearly and legibly, all these Cover Texts: Front -Cover Texts on the front cover, and Back-Cover Texts on the back cover. Both covers must also clearly and legibly identify you as the publisher of these copies. The front cover must present the full title with all words of the title equally prominent and visible. You may add other material on the covers in addition. Copying with changes limited to the covers, as long as they preserve the title of the Document and s atisfy these conditions, can be treated as verbatim copying in other respects. If the required texts for either cover are too voluminous to fit legibly, you should put the first ones listed (as many as fit reasonably) on the actual cover, and continue the rest onto adjacent pages. If you publish or distribute Opaque copies of the Document numbering more than 100, you must either include a machine-readable Transparent copy along with each Opaque copy, or state in or with each Opaque copy a computer -network location from which the general network -using public has access to download using public -standard network protocols a complete Transparent copy of the Document, free of added material. If you use the latter option, you must take reasonably prudent steps, when you begin distribution of Opaque copies in quantity, to ensure that this Transparent copy will remain thus accessible at the stated location until at least one year after the last time you distribute an Opaque copy (directly or through your agents or retailers) of that edition to the public. It is requested, but not required, that you contact the authors of the Document well before redistributing any large number of copies, to give them a chance to provide you with an updated version of the Document.

4. MODIFICATIONS

You may copy and distribute a Modified Version of the Document under the conditions of sections 2 and 3 above, provided that you release the Modified Version under precisely this License, with the Modified Version filling the role of the Do cument, thus licensing distribution and modification of the Modified Version to whoever possesses a copy of it. In addition, you must do these things in the Modified Version: A. Use in the Title Page (and on the covers, if any) a title distinct from that o f the Document, and from those of previous versions (which should, if there were any, be listed in the History section of the Document). You may use the same title as a previous version if the original publisher of that version gives permission. B. List on the Title Page, as authors, one or more persons or entities responsible for authorship of the modifications in the Modified Version, together with at least five of the principal authors of the Document (all of its principal authors, if it has fewer than f ive), unless they release you from this requirement. C. State on the Title page the name of the publisher of the Modified Version, as the publisher. D. Preserve all the copyright notices of the Document. E. Add an appropriate copyright notice for your modi fications adjacent to the other copyright notices. F. Include, immediately after the copyright notices, a license notice giving the public permission to use the Modified Version under the terms of this License, in the form shown in the Addendum below. G. Preserve in that license notice the full lists of Invariant Sections and required Cover Texts given in the Document's license notice. H. Include an unaltered copy of this License. I. Preserve the section Entitled "History", Preserve its Title, and add to it an item stating at least the title, year, new authors, and publisher of the Modified Version as given on the Title Page. If there is no section Entitled "History" in the Document, create one stating the title, year, authors, and publisher of the Document as given on its Title Page, then add an item describing the Modified Version as stated in the previous sentence. J. Preserve the network location, if any, given in the Document for public access to a Transparent copy of the Document, and likewise the netwo rk locations given in the Document for previous versions it was based on. These may be placed in the "History" section. You may omit a network location for a work that was published at least four years before the Document itself, or if the original publish er of the version it refers to gives permission. K. For any section Entitled "Acknowledgements" or "Dedications", Preserve the Title of the section, and preserve in the section all the substance and tone of each of the contributor acknowledgements and/or d edications given therein. L. Preserve all the Invariant Sections of the Document, unaltered in their text and in their titles. Section numbers or the equivalent are not considered part of the section titles. M. Delete any section Entitled "Endorsements". S uch a section may not be included in the Modified Version. N. Do not retitle any existing section to be Entitled "Endorsements" or to conflict in title with any Invariant Section. O. Preserve any Warranty Disclaimers. If the Modified Version includes new f ront-matter sections or appendices that qualify as Secondary Sections and contain no material copied from the Document, you may at your option designate some or all of these sections as invariant. To do this, add their titles to the list of Invariant Secti ons in the Modified Version's license notice. These titles must be distinct from any other section titles. You may add a section Entitled "Endorsements", provided it contains nothing but endorsements of your Modified Version by various parties --for example , statements of peer review or that the text has been approved by an organization as the authoritative definition of a standard. You may add a passage of up to five words as a Front -Cover Text, and a passage of up to 25 words as a Back -Cover Text, to the end of the list of Cover Texts in the Modified Version. Only one passage of Front-Cover Text and one of Back-Cover Text may be added by (or through arrangements made by) any one entity. If the Document already includes a cover text for the same cover, previously added by you or by arrangement made by the same entity you are acting on behalf of, you may not add another; but you may replace the old one, on explicit permission from the previous publisher that added the old one. The author(s) and publisher(s) of the Document do not by this License give permission to use their names for publicity for or to assert or imply endorsement of any Modified Version.

5. COMBINING DOCUMENTS

You may combine the Document with other documents released under this License, under the terms defined in section 4 above for modified versions, provided that you include in the combination all of the Invariant Sections of all of the original documents, unmodified, and list them all as Invariant Sections of your combined work in its lic ense notice, and that you preserve all their Warranty Disclaimers. The combined work need only contain one copy of this License, and multiple identical Invariant Sections may be replaced with a single copy. If there are multiple Invariant Sections with the same name but different contents, make the title of each such section unique by adding at the end of it, in parentheses, the name of the original author or publisher of that section if known, or else a unique number. Make the same adjustment to the sectio n titles in the list of Invariant Sections in the license notice of the combined work. In the combination, you must combine any sections Entitled "History" in the various original documents, forming one section Entitled "History"; likewise combine any sect ions Entitled "Acknowledgements", and any sections Entitled "Dedications". You must delete all sections Entitled "Endorsements".

6. COLLECTIONS OF DOCUMENTS

You may make a collection consisting of the Document and other documents released under this Lice nse, and replace the individual copies of this License in the various documents with a single copy that is included in the collection, provided that you follow the rules of this License for verbatim copying of each of the documents in all other respects. Y ou may extract a single document from such a collection, and distribute it individually under this License, provided you insert a copy of this License into the extracted document, and follow this License in all other respects regarding verbatim copying of that document.

7. AGGREGATION WITH INDEPENDENT WORKS

A compilation of the Document or its derivatives with other separate and independent documents or works, in or on a volume of a storage or distribution medium, is called an "aggregate" if the copyright resulting from the compilation is not used to limit the legal rights of the compilation's users beyond what the individual works permit. When the Document is included in an aggregate, this License does not apply to the other works in the aggregate which are not themselves derivative works of the Document. If the Cover Text requirement of section 3 is applicable to these copies of the Document, then if the Document is less than one half of the entire aggregate, the Document's Cover Texts may be placed on covers that bracket the Document within the aggregate, or the electronic equivalent of covers if the Document is in electronic form. Otherwise they must appear on printed covers that bracket the whole aggregate.

8. TRANSLATION

Translation is considered a k ind of modification, so you may distribute translations of the Document under the terms of section 4. Replacing Invariant Sections with translations requires special permission from their copyright holders, but you may include translations of some or all I nvariant Sections in addition to the original versions of these Invariant Sections. You may include a translation of this License, and all the license notices in the Document, and any Warranty Disclaimers, provided that you also include the original Englis h version of this License and the original versions of those notices and disclaimers. In case of a disagreement between the translation and the original version of this License or a notice or disclaimer, the original version will prevail. If a section in t he Document is Entitled "Acknowledgements", "Dedications", or "History", the requirement (section 4) to Preserve its Title (section 1) will typically require changing the actual title.

9. TERMINATION

You may not copy, modify, sublicense, or distribute th e Document except as expressly provided for under this License. Any other attempt to copy, modify, sublicense or distribute the Document is void, and will automatically terminate your rights under this License. However, parties who have received copies, or rights, from you under this License will not have their licenses terminated so long as such parties remain in full compliance.

10. FUTURE REVISIONS OF THIS LICENSE

The Free Software Foundation may publish new, revised versions of the GNU Free Documentat ion License from time to time. Such new versions will be similar in spirit to the present version, but may differ in detail to address new problems or concerns. See http://www.gnu.org/copyleft/. Each version of the License is given a distinguishing version number. If the Document specifies that a particular numbered version of this License "or any later version" applies to it, you have the option of following the terms and conditions either of that specified version or of any later version that has been pub lished (not as a draft) by the Free Software Foundation. If the Document does not specify a version number of this License, you may choose any version ever published (not as a draft) by the Free Software Foundation. ADDENDUM: How to use this License for yo ur documents To use this License in a document you have written, include a copy of the License in the document and put the following copyright and license notices just after the title page: Copyright (c) YEAR YOUR NAME. Permission is granted to copy, distr ibute and/or modify this document under the terms of the GNU Free Documentation License, Version 1.2 or any later version published by the Free Software Foundation; with no Invariant Sections, no Front -Cover Texts, and no Back-Cover Texts. A copy of the li cense is included in the section entitled "GNU Free Documentation License". If you have Invariant Sections, Front -Cover Texts and Back -Cover Texts, replace the "with...Texts." line with this: with the Invariant Sections being LIST THEIR TITLES, with the Fr ont-Cover Texts being LIST, and with the Back -Cover Texts being LIST. If you have Invariant Sections without Cover Texts, or some other combination of the three, merge those two alternatives to suit the situation. If your document contains nontrivial examp les of program code, we recommend releasing these examples in parallel under your choice of free software license, such as the GNU General Public License, to permit their use in free software.

CPSIA information can be obtained at www.ICGtesting.com
Printed in the USA
LVOW041803231012
304103LV00003B/134/P